Island Colonization

New or recently sterilized islands (for example through volcanic activity) provide ecologists with natural experiments in which to study colonization, development and establishment of new biological communities. Studies carried out on islands like this have provided answers to fundamental questions as to what general principles are involved in the ecology of communities and what processes underlie and maintain the basic structure of ecosystems. These studies are vital for conservation biology, especially when evolutionary processes need to be maintained in systems in order to maintain biodiversity. The major themes are how animal and plant communities establish, particularly on 'new land' or following extirpations by volcanic activity. This book comprises a broad review of island colonization, bringing together succession models and general principles, case studies in which Professor Ian Thornton was intimately involved, and a synthesis of ideas, concluding with a look to the future for similar studies.

Professor IAN THORNTON (La Trobe University, Melbourne) who was one of the world's leading island biogeographers, working primarily on the volcanic islands of the Pacific, died in 2002. Amongst the papers he left behind was the manuscript of this book, with an express wish that should anything happen to him, his friend and colleague, Professor TIM NEW (also La Trobe University) would finish the book for him, which he has done.

Island Colonization

The Origin and Development of Island Communities

IAN THORNTON
Formerly of La Trobe University, Melbourne

Edited by

TIM NEW
La Trobe University, Melbourne

CAMBRIDGE
UNIVERSITY PRESS

CAMBRIDGE UNIVERSITY PRESS
Cambridge, New York, Melbourne, Madrid, Cape Town, Singapore, São Paulo

Cambridge University Press
The Edinburgh Building, Cambridge CB2 8RU, UK

Published in the United States of America by Cambridge University Press, New York

www.cambridge.org
Information on this title: www.cambridge.org/9780521854849

First published 2007

Printed in the United Kingdom at the University Press, Cambridge

A catalogue record for this publication is available from the British Library

ISBN-13 978-0-521-85484-9 hardback
ISBN-13 978-0-521-67106-4 paperback

Contents

Editorial preface

Ian Walter Boothroyd Thornton, Emeritus Professor of Zoology at La Trobe University, Melbourne, died in Bangkok on 1 October 2002, on his way home from Laos, where he had been advising the Laos National University on science course development. Among his papers he left an incomplete initial draft of a book he had long planned to write on the development of island communities – an interest he had pursued vigorously for much of his academic life. Ian had discussed aspects of his book with me, and other colleagues, and had expressed the wish to his wife, Ann, that should anything happen to prevent him from finishing it himself, I might be able to complete the work and see it through to publication. As anyone who has attempted any similar task will know, such an exercise is daunting in both scope and responsibility! Although we had debated a number of the themes included, and I had participated in extensive field expeditions (including four trips to Krakatau) with Ian over some 30 years, our perspectives sometimes differed considerably.

The dilemmas of editing a close colleague and friend's posthumous work have been summarized admirably (and, coincidentally, in the autobiography of another distinguished Pacific region entomologist, Robert Usinger), and I have tried to follow the spirit of the perspective given there by Gorton Linsley and Lin Gressitt (1972). They noted that their subject could have no opportunity to read and edit the final manuscript, and that they wished to retain his words in their original form. They therefore opted to limit editorial effort to verification of facts, correction of errors and obvious 'lapsi' (*sic*), but otherwise simply sought to improve the format whilst preserving the original text. Most of the words in this book are indeed Ian's, and some chapters have been changed little from his files. More revision and augmentation has been needed for others, together with illustrations and tables (as guided by the text), some later material and references. I have attempted to make my limited additions both seamless and in accord with Thornton's views.

Reviewers of an early draft suggested that the then extensive 'cultural content' of the book should be reduced somewhat, to increase its primary relevance to ecology. I have done this, but retained some flavour of the effects of volcanism on humanity. However, much of this aspect has been dealt with in a very readable recent account by de Boer and Sanders (2002), so that the intriguing questions of the wider influences of the Theran and Krakatau eruptions that Thornton had treated extensively in his draft are now largely redundant. Winchester's (2003) book on Krakatau also provides much background on broader effects on people. Again from reviewers, I was asked to consider two options for how the work might be developed. First, to present it with limited augmentation as Ian Thornton's creation and a summary of his thoughts on some important ecological themes as developed during a long period of practical investigation. Second, to modify and enlarge it far more extensively as a co-author, a path that one reviewer considered to be more valuable. I opted for the first of these, in the belief that further development would introduce the likelihood of submerging the personal perspective that is the core of the book, and lessen Thornton's impact by introducing alternative views: analytical biogeography is a controversial topic! For the same reason, I have included only minimal references to papers published after 2002.

The book comprises several distinct parts, each of several chapters. The first part introduces the themes discussed and provides some of the historical and theoretical background that biologists have used to study island communities. Part II deals with the processes of recolonization after devastation by volcanic activity, with Part III following this with appraisals of specific islands, some of which have been studied in considerable detail. The later trends of assembly from these foundation communities are treated in Part IV, with broader integration (using the same cases) presented in Part V. Finally, Ian looks forward from this discourse to endorse and evaluate the importance of such studies in understanding and sustaining Earth's biota. This is not a textbook. Rather, it presents a chosen suite of ideas, perspective and reasoning by a leading Pacific-region biogeographer, together with diversions on the wider effects of volcanic eruptions on humanity, and how both natural communities and human civilization may develop subsequently. It is founded in the ideas and conflicts of island biogeography, and the biological and other information relevant to Ian's themes is documented succinctly. The final 'message', of the importance of understanding community development in order to conserve both natural communities and the underlying processes that sustain them, is one with which few biologists would disagree. He also raises the important topic of 'what we don't know', and ecologists seeking fruitful ideas will find much to stimulate their minds, and to challenge their ingenuity.

Ian Thornton was passionate about islands: a true 'islomaniac' in Schoener's (1988) sense of having 'a powerful attraction to islands'; and his enthusiasm was infectious. Early in his academic career, he spent a sabbatical period in the Hawaiian Islands, where his studies on the remarkable endemic radiations of Psocoptera of the archipelago illuminated patterns of speciation on such isolated islands. Later, he pursued similar studies in the Galapagos, and a book resulting from that trip (Thornton 1971) is a classic in the literature of that intriguing archipelago. However, these early studies founded in post-colonization events on islands and clarifying patterns of insect evolution and distribution whetted Ian's appetite for larger questions, leading him to investigate and consider the more general origins of island faunas and the extents to which their communities are structured or 'determined' in some way.

In many ways, and as noted above, this is a very personal book. Following from his imposing synthesis of Krakatau (Thornton 1996a) and later studies on Long Island (resulting from a strenuous expedition he organized and led whilst in his 70s: Thornton 2001), he brings together his experiences and previous studies, together with those of a number of his friends and colleagues, to address issues of fundamental evolutionary and biogeographical interest. Thus, John Edwards (whose studies on Mount St Helens are discussed here) visited Krakatau with Ian, and was his major collaborator on Long Island; Sturla Fridriksson (whose magisterial studies of Surtsey Ian recognized as one of the closest parallels to Anak Krakatau) also visited Krakatau and Ian, Surtsey. Rob Whittaker's group at Oxford and our group at La Trobe enjoyed much intellectual controversy over our mutual interests in Krakatau – and Ian revelled in controversy! He continually sought the intellectual input of colleagues versed in those fields of biology with which he was less familiar; he enjoyed encouraging young people, and many recent graduates and graduate students were enthusiastic participants in his expeditions – which were pivotal experiences in their development. As Ian wrote in his acknowledgments to *Krakatau*: 'Although my Indonesian colleagues flatter me with the saying "naga tua, tenada muda" ("old dragon, young power"), in recent years the need for auxiliary, really young power has increased.' However, his enthusiasms and drive never diminished.

Ian would, I suspect, have made his major acknowledgement in this book to Ann, with whom he first viewed Krakatau in 1982 (a sighting that was to prove pivotal in focusing his studies on island community development) and to whom he dedicated his book on Krakatau. I would echo this acknowledgment in thanking Ann for making his draft available, and for entrusting me with trying to bring it to fruition. Some of his many colleagues and correspondents, participants or collaborators in expeditions, sounding boards for Ian's speculations, or critics of his written drafts, are cited in context in the book and it would be presumptuous of me to list them here in any order of priority. However, a list of his major co-authors and collaborators from the expeditions and suites of papers from recent expeditions provide a partial summary of the debts he would assuredly acknowledge. I regret that these thanks must be incomplete: the accumulation of a lifetime debt of acknowledgement is not easy for anyone else to appraise properly. But, many times in meeting Ian in the departmental corridor up to a decade after his formal (but nominal) retirement from the Department of Zoology, he would make some comment along the lines of

'I've just heard from so-and-so. He/she reckons that . . .. What do you think?' The unconscious assimilation of such inputs is part of the intellectual development of any scientist; but the depth and breadth of Ian Thornton's interests and the generosity with which he would respond with constructive (and, often, original) comment was typical of the man, and of the regard in which his scientific work is held.

Tim New
Melbourne, February 2006

Biographical memoir

New, T. R., Smithers, C. N. and Marshall, A. T. (2005). Ian Walter Boothroyd Thornton 1926–2002. *Historical Records of Australian Science* **16**, 91–106.

Acknowledgements

The following editors and publishers are thanked for granting permission to reproduce or modify tabular or illustrative material: Blackwell Science, Oxford; Ecological Society of America, Washington DC; Elsevier Science, Oxford; The Royal Society, London; Springer Science and Business Media, Heidelberg; Professor Junichi Kojima (*Japanese Journal of Entomology*); Professors Osamu Tadauchi and Junichi Yukawa (*Esakia*). Every effort has been made to obtain permissions for such use, and the publishers would welcome news of any unintended oversights.

I would also like to thank Professor Hideo Tagawa, Kagoshima University, Japan for his invaluable compilation of the 46 scientific papers resulting from his team's extensive research on the Krakatau archipelago (Tagawa 2005), bringing together papers that were previously highly scattered and, some, difficult to access. Dr Borgthór Magnússon has generously sent me recent papers on Surtsey.

I appreciate very much the enthusiasm of Ward Cooper, through whom this book was submitted to Cambridge University Press, and the constructive comments made by reviewers of that earlier draft. His successor at Cambridge, Dominic Lewis, has been very patient in awaiting the completed book.

The impeccable copy-editing by Anna Hodson has enhanced the quality of the manuscript markedly, and I greatly appreciate her perceptive comments and advice.

The cover photograph, of Anak Krakatau erupting in 1993, is by Igan S. Sutawidjaja.

PART I

Theoretical and experimental studies

CHAPTER ONE

Introduction

It is clear that at present the peoples of this planet, in general, do not treat it as a single superhabitat for all its life forms, including its human population. There is little evidence that even the so-called developed countries have this global philosophy. The reluctance of nations such as the USA and Australia to come to terms with clear danger signs of global warming is a striking example of the priorities adopted by 'developed' nation states in the third millennium. They are unwilling to forfeit some economic well-being by taking precautionary steps that would reduce the rate of human-induced climate change. The argument that the change has not been proved beyond doubt to the satisfaction of the political leaders is a shallow excuse. So widespread are the repercussions of climatic change, and so potentially detrimental to human welfare generally, that the safe global strategy would be to begin taking preventative steps now, and argue about the validity of scientists' warnings later. But the bottom lines of national economies take precedence over global concerns, and in democracies these choices must reflect the wishes of the majority, or governments that espoused such approaches would fall. In the end, the people of the 'leading countries' of the world must bear the responsibility.

But I am an optimist. I believe that the overriding philosophy within the next century will be framed within a truly global outlook. More and more countries will come to regard their plant and animal species, and the ecosystems that they help to form, as part of their nation's and the world's natural wealth. A few, notable among them such small countries as Iceland and Costa Rica, already appear to do so. Perhaps at the time that a world government is accepted, and possibly as a prelude to that development, considerations of first continental, and then global, ecology will gradually override nationalistic economic rationalism. People generally may then come to regard the degree of disturbance to Earth's ecosystems that is involved in the exploitation of some of its resources as an unacceptable price to pay for the use of them. Until then, it is likely that the large-scale, rapid consumption and destruction of some of Earth's non-renewable or slowly renewable resources will continue, and at an accelerating rate. This rapid destruction of resources is already endangering complex ecosystems such as tropical forests, about which we still know very little, and which may yet be found to harbour untold wealth, for example in the form of biotic components with economic and medical applications.

Upon the future dawning of what may be viewed as the Age of Global Management, mankind may have to intervene on a large scale in order to maintain some natural ecosystems at levels that permit the normal functioning of global ecological processes. Not only will those ecosystems that have survived need to be conserved, but those that have been seriously damaged or destroyed may have to be repaired or restored. The large-scale restoration of complex ecosystems such as tropical rainforest would be an enormous challenge to our descendants, but one that would be accepted by ecologists if they were given supranational political support. Any such undertakings would need to be as accurate restorations of the natural systems as possible; mere rehabilitation, such as is now often practised following relatively small-scale industrial, mining or agroforestry disturbance, would not suffice. Daniel Janzen and his team of biologists in Mesoamerica have already made a start in the large-scale rehabilitation of a tropical forest ecosystem. They are the pioneers, and their project will surely be followed by others in the decades and centuries to come.

Crucial to any attempts to restore, or re-create a natural ecosystem, will be some knowledge of its 'embryology', how it has arisen and developed naturally. What might be called the science of 'developmental ecology' would combine some of the rapidly fledging applied subsciences of restoration and rehabilitation ecology with much of present community ecology and island biogeography, including studies on the origins and development of ecosystems and communities. For this last, it is obvious that study ecosystems are needed that are at or near the beginning of their development, so that we may follow, and attempt to understand, the natural processes of species' colonization and biotic assembly and succession that characterize the making of functioning communities.

These study systems may be theoretical models, of course, or experimental systems in the laboratory or in the field, and there have already been studies along these lines, notably in the United States. Rarely, however, geological events provide us with large-scale natural situations that involve the creation of new ecosystems. These events usually involve the production of a *tabula rasa*, or 'clean slate', by total devastation of a previous ecosystem by some natural catastrophe. More rarely, they involve the creation of a completely new, pristine substrate, on which an ecosystem is assembled de novo, from scratch.

The processes involved in ecosystem and community assembly are likely to be shown most clearly when the new system is distinctly separated from its source, or reservoir of potential participants, so that colonizing events and their ecological repercussions are clearly recognizable. This means that work will focus on the study of isolates, whether they be real or simulated, ecological or geographical, natural or artificial.

Theoretical, model systems have the advantage that certain basic assumptions may be incorporated into the model, and the consequences of changing these or various parameters of the developmental process can be studied in computer simulations. But convincing models must be founded in biological reality rather than solely in conceptual perspectives.

'Real' island colonization situations can be created in the laboratory, and the effects of changing various factors, in terms of the composition of the resultant community, can be determined – a subscience that may be called 'experimental developmental ecology'. For practical convenience such experimental systems have usually involved the controlled introduction (both rate and sequence may be altered) of freshwater bacteria, algae, protozoans and other microbiota to artificial aquatic island microcosms on the laboratory bench.

Experimental work that is necessarily less precise but closer to the natural situation than laboratory studies involves the colonization of artificial substrate 'islands' in natural conditions. Some early pioneer studies of this kind have been the monitoring of colonization by aquatic microbiota of glass slides placed in streams (Patrick 1967), the build-up of aquatic communities in dishes of sterile water placed in natural situations (Maguire 1963), and the growth of populations of settling ('fouling') communities of marine organisms on settlement plates placed in the sea (Osman 1982). Examples of isolated physical habitats or critical resources (such as the isolated hostplants needed by insect specialist herbivores in diverse stands of vegetation) could be multiplied ad nauseam. Many have acknowledged value in disciplines such as biological control, in which an 'island plant' or 'island host' may be exploited by consumers. They collectively demonstrate the need for many organisms to continually disperse and colonize, in order to 'track' their needs as conditions change. The term 'island habitats' was used, for example, by Beaver (1977) in referring to dead snails, sought by the carrion-feeding flies which use them as breeding sites. As for more conventional islands, such substrates or discrete units can be the foundation for community development, with the course of that development subject to processes such as succession and stochastic events (such as the snail carcass being eaten).

As another example, 'phytotelmata' is the collective name for natural plant-held water bodies, such as those in hollows at the junction of branches, in the axils of leaves or in tree holes. These very small natural rainwater pools, of course, are habitat islands for their inhabitants. Individual phytotelmata support relatively simple, isolated microcosms, and these are available in multiplicate (Kitching 2000). Such natural mini-islands are study units that have much to offer the student of island biotas, not least because they are sufficiently abundant for experimental manipulations to be made. Thus, variations and deviations from the 'standard microcosm' can be interpreted and, to some extent, explained in terms of community changes. Moreover, modelling approaches to these studies, such as that of Post and Pimm (1983), have relevance not only to community ecologists studying water-filled tree holes, but also to island ecologists and biogeographers.

Natural restoration may be studied on land on a larger scale following the devastation of an ecosystem by purposeful or accidental human activity, such as clear-felling or fire. More rarely, nature herself provides the whole 'experiment', including the 'new' substrate, for example following natural events such as fire, hurricane, flood, glacial action or volcanic activity involving lava flows or tephra falls. Such cases, however, are not always isolated from adjacent areas that have been spared destruction, from which colonists may arise. They are thus less instructive, in the context of this book, than cases that involve islands. Moreover, the destruction is usually incomplete. Almost invariably, a 'foundation community' (usually incompletely documented) is already present – soil biota, surviving plants and others – and the processes then involve aspects of succession, with the trajectories determined in part by these source communities. The base-line from which restoration of a community or the assembly of a new community must proceed is thus not absolutely clear or uniform; recolonization can proceed from remnant foci of survivors, and often does so. To a large extent each such case is unique in detail, although the underlying principles driving change may be more widespread.

Very occasionally, biologists have the chance to follow the natural assembly of an isolated ecosystem on a macro-scale, and from a base-line that is known to be zero or very near to zero. Isolated sterile substrates, or at least substrates from which all macrobiota (animals and plants) have been extirpated, may be formed by the destruction of an existing island community by a natural cataclysm. For example, a terrain that has been scraped clean by the slow, inexorable action of a glacier may be exposed when the glacier retreats, providing a new substrate, almost literally a cleaned slate, on which new communities may now develop. Or following explosive volcanic activity, the subsequent reassembly of a community may be monitored by periodic biological surveys. Where the volcano happens to be an island, the monitoring is facilitated because the sources of organisms are discrete from it and colonists more easily identified.

Perhaps the best natural situation of all for the study of the origins of ecosystems is the emergence of a new volcanic island from the sea or from the crater lake of an existing volcano. Such an event provides a rare opportunity to monitor the assembly of a terrestrial community on a substrate that is undoubtedly originally devoid of terrestrial life, and, moreover, clearly separated from possible sources of colonists. From tree holes to emergent volcanic islands may seem like a giant leap, but island biogeographers need to keep a wide perspective. In seeking to understand the processes that operate in the colonization of 'real islands', of whatever size, we need to be aware of, and where appropriate make connection with, the work of theoreticians and of both laboratory and field community ecologists.

However, an important point to appreciate in seeking analogy is that laboratory island studies can provide relatively rigorous and controlled information on the basis of adequate replication and statistical analyses, whereas real world islands are each unique. We may seek, and to some extent reveal, parallels between different islands, but each is a single unreplicated experiment in ecological development.

Types of islands
Continental islands

Unlike an oceanic island, which emerged from the sea as an isolated land mass, usually by volcanic activity, and has received all its biota from outside sources, a continental island is formed by the isolation of a piece of land that was formerly part of a larger land mass. Many continental islands were formed as a result of rising sea levels following lowered levels in the Pleistocene glaciations, when much of Earth's water was locked up in the form of ice. The rocks of continental islands may be sedimentary, metamorphic or igneous, or any combination of these, but are similar to those of their parent mainland. In some cases the island is separated by a shallow narrow strait, such as the separation of Bali from Java in Indonesia; in other cases, such as Fiji or the Seychelles, the continental shelf may be so extensive and so far below sea level (perhaps 200 m) that the islands appear to be oceanic. At the time of their creation by separation from the larger land mass, continental islands contained a portion of the latter's biota. Over time, the diversity (species richness) of the newly created fragment usually declines as an adjustment to what is now a smaller and isolated environment.

Ecological 'continental islands' may also be formed naturally. For example, when lava flows around a piece of higher ground, often round the same piece of ground in successive eruptions, the biota of the isolated area, or *kipuka*, is spared destruction for a long period and becomes surrounded by a sea of barren lava or, at least, by a community at a very much younger stage of development. Mountain tops also become 'continental' ecological islands when climatic change changes the tree-line, leaving areas of montane biota as ecological islands, surrounded by lower lands within which many mountain organisms cannot survive and through which they cannot pass. Many such areas thus support taxa long isolated from their closest relatives on other, sometimes nearby, mountain peaks. As Carlquist (1965) elegantly put it, 'Mountain tops are like islands in the upper air', which he contrasted with the deep ocean trenches as islands for organisms coping with very different extreme environments. Lakes, geographical islands in reverse, may be 'continental' if they are formed by the entrapping and isolation of part of a once more extensive water body.

Continental islands may also be formed by human activity. Every piece of remnant woodland left in an agricultural landscape, and every natural reserve preserved in an urban area, is a 'continental' ecological island of natural vegetation surrounded by a sea of modified habitat. Two such natural reserves in urban areas that have been studied in this context are the famous 84-ha Botanical Gardens in the town of Bogor, Java (Diamond *et al.* 1987) and the 400-ha Kings Park Reserve in the centre of Perth, Australia (Recher and Serventy 1991). The 72-ha Bukit Timah rainforest reserve on Singapore Island and Barro Colorado Island, Panama, are other well-known examples of continental islands formed as a result of human activity. Barro Colorado Island, a former hilltop, was isolated early in the twentieth century by the construction of the Panama Canal.

Oceanic islands

Oceanic islands, in contrast, are those that have originated from a lake or from the ocean bed, usually as a result of subaquatic volcanic activity, and were never connected to a mainland. They are usually, but not necessarily, extremely isolated. They may be the result of submarine volcanic activity at a geological 'hot spot' in the middle of a tectonic plate (e.g. Hawaii, Fig. 1.1), at the mid-ocean

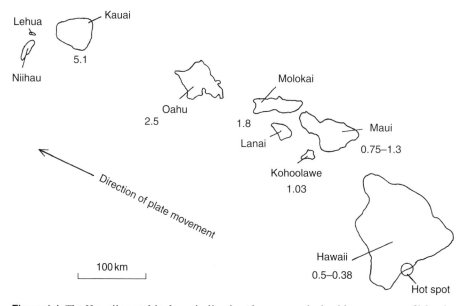

Figure 1.1 The Hawaiian archipelago, indicating the progressively older sequence of islands as the Pacific tectonic plate moves westward over the hot spot currently near the southeastern part of Hawaii. No other oceanic archipelago supports such an array of endemic species, and studies of major radiations of insects (such as drosophilid flies) have done much to illuminate patterns of speciation in such isolated environments. The islands shown are the most recent (ages shown in millions of years) in an island chain that extends for some 5800 km across the Pacific.

ridge where new plate material is being formed (Iceland, Tristan da Cunha), at the junction of two or more ocean ridges (Galapagos), or beyond a subduction zone, where the movement of the subducting plate results in magma reaching the surface usually along an arc (e.g. the Solomon Islands). Subsidence and erosion may transform oceanic islands into low atolls (e.g. Wake), or subsequent uplift may result in 'raised atolls' of limestone (like the Lau group of the Fiji archipelago), but originally they were all high islands with volcanic mountains. Very occasionally, a freshwater lake may arise de novo, as when a volcanic caldera fills with rainwater. From the point of view of aquatic organisms a new island has been formed, and, by our definition, an 'oceanic' one.

There are a number of examples of very young oceanic islands that are not particularly isolated. Anak Krakatau emerged initially from Krakatau's submarine caldera between Java and Sumatra on the Sunda island arc in 1930. Surtsey emerged in 1963 from the subatlantic mid-ocean ridge, off Iceland. The new Fijian island, Lomu, named after a famous rugby player, emerged in 1994. Motmot is an islet that emerged, probably in the late 1950s, from the freshwater caldera lake of Long Island. Rakaia Island was formed in 1878 when the Vulcan volcano erupted in the Rabaul caldera, New Britain, and became joined to the mainland during a subsequent eruption in 1937. The nascent island Loihi, off the east coast of the big island of Hawaii, is still developing below sea level – as an oceanic island not yet 'born'.

Island biogeographers have studied all these types of islands and their biotas. The ways in which their ecological communities have developed, and the pathways of their assembly and evolution, are of particular interest on oceanic islands. There, they illustrate processes that are of much wider importance in seeking to understand the dynamics and restoration of natural ecosystems. Together, these studies contribute much to our understanding of change in the natural world. This book is an attempt to explore and integrate studies on oceanic and continental islands, and to see what features, if any, are common to the growth and development of different kinds of island communities.

CHAPTER TWO

Theoretical and experimental colonization

Theoretical models

The colonization of islands can be studied in theory by constructing models on the computer. Such models require that certain assumptions be built into the model, and, although these are necessarily simplified (and always fail to take into account all the parameters relevant to natural situations), insights can be gained from their use. As an example, one of the questions that can be investigated in this way is the relative roles of chance (stochasticity) and determinism in the colonization process. This question has been debated vigorously by island biogeographers, and an understanding of the different points of view is important in introducing the topics.

Chance and determinism

In the process of assembly of a functioning community from a number of species of living things, both stochastic and deterministic elements are almost always involved. Several biologists have attempted to assess the interplay of these two elements of the process. After using a very simplified model, Seamus Ward and I concluded that, on theoretical grounds, in the early stages of colonization stochasticity (in the sense of the likelihood that in a population of similar islands the communities that are assembled will be different from one another) will be low. Pioneer colonizers are invariably species with good dispersal powers and/or establishment characteristics. Their arrival rates will be high and very similar to one another, and the intervals between colonizing events (colonizing intervals) will be short. Also, later in the colonization process, when arrival rates and colonizing intervals of the species concerned are considerably more variable, as with pioneers versus late-successional species, stochasticity will again be low.

Still later, when the colonizing species have fairly similar, but low, arrival rates (for example, pairs of species on distant islands), priority (the order of arrival) will vary a great deal between cases, and because the presence of the first arrival may enhance or decrease the chances of establishment of later arrivals, the particular sequence of arrival will greatly influence the final community outcome. At this stage, stochasticity will be high (Ward and Thornton 2000).

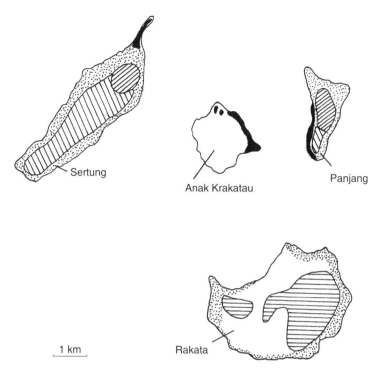

Figure 2.1 The islands of the present Krakatau archipelago, indicating the major forest categories present on each island. Black, *Casuarina*; horizontal hatch, *Neonauclea*; dots, *Terminalia*, coastal; left–right diagonal, *Timonius*; right–left diagonal, *Dysoxylum*; blank, none (Anak Krakatau) or unexplored (Rakata). (After Whittaker *et al.* 1989, as modified by Thornton 1992a.)

When the colonizing species differ considerably, but not extremely, in dispersal ability, our model showed that stochasticity may still be high, but variation in colonizing interval, rather than variation in priority, may be the main contributor to the stochasticity (Ward and Thornton 1998). This theoretical finding is supported by the known course of colonization of the Krakatau islands by plants.

The three older Krakatau islands (Fig. 2.1), some 4 km apart, have been recolonized by plants and animals since Krakatau's great devastating eruption of 1883. The course of development of mixed forest on the islands diverged after the 1920s, so that on two of them, Panjang and Sertung, but not on the third, Rakata, areas of forest were dominated by one of two animal-dispersed trees, *Timonius compressicaulis* and *Dysoxylum gaudichaudianum*. On Rakata, however, the forest came to be dominated by the wind-dispersed tree *Neonauclea calycina*, which has a dispersal rate higher by perhaps two orders of magnitude than those of the two zoochorous species.

On Rakata the colonizing intervals between *Neonauclea* and *Timonius* and between *Neonauclea* and *Dysoxylum* were at least 24 and 34 years respectively (Table 2.1), but on Panjang, where *Timonius* and *Dysoxylum* became the dominants, the corresponding intervals were zero and at most 3 years, and *N. calycina* forest was not able to develop when its competitors followed so quickly.

Table 2.1 *Ranges of possible arrival dates of the major canopy tree species on islands of the Krakatau archipelago. Earliest possible arrival date is the year following the last botanical survey in which the species was not found; the second date is that of the first survey to detect the species*

	Rakata	Panjang	Sertung	Anak Krakatau
Neonauclea calycina	1898–1905	1909–29	1909–20	1950–51[a],1972–79
Timonius compressicaulis	1919–29	1909–29	1930–82	1980–81
Dysoxylum gaudichaudianum	1952–79	1909–32	1930–82	1980–81

[a] Extirpated by volcanic activity
Source: Ward and Thornton (2000).

Although here dominance, rather than exclusion, is the effective outcome, the pattern appears to support conclusions from the model and illustrates the highly stochastic effect of colonizing interval when relatively poor dispersers are involved. The observed outcome is not at variance with the model's prediction that variable colonizing interval, rather than varying priority, provided stochasticity in this case.

Because in the forest-forming stage of the succession there is likely to be more variability both in priority and in colonizing interval, stochasticity is likely to be higher than it was in the early pioneer phase. As forest formation on the Krakataus proceeded, however, and the canopy closed, a 'second wave' of aerially dispersed species, including many shade-loving ferns, colonized the new mesic habitats that were created by forest closure. Presumably the component species of this wave were determined largely by high dispersal powers and the constraints on establishment provided by the new, more mesic environment of the forested island. The last colonizing wave comprised largely zoochorous trees, and the positive feedback associated with dispersal by means of fruit-eating birds and bats led to a steady rise in the importance of this wave. Immigration rates are relatively low and similar for all potential colonists in this wave and, as for *Dysoxylum* and *Timonius*, both priority and colonizing interval should provide high stochasticity, resulting in variation between cases.

Experimental islands in the laboratory

In 1972, Robert MacArthur had suggested that differences in invasion sequence may influence the course of colonization and even the final equilibrium number of species. He thought that an island may in fact have a number of different colonization curves, reflecting different sequences of colonization from the same 'mainland pool'. In the 1980s and 1990s, American experimental community ecologists followed up and extended MacArthur's idea. They developed techniques for creating miniature freshwater 'habitat islands' in the laboratory, colonization situations in which the immigration of species to the system could be controlled and manipulated. Dickerson, Robinson and Edgemon, for example, assembled microcosmic freshwater island communities by the addition, one at a time, of species of microscopic freshwater organisms such as ciliate protozoans, bacteria, algae and rotifers. They could change the order of immigration from their arbitrary 'mainland pool' of species, of course (for example, they could reverse it), or, keeping the order the same, they could alter the intervals between 'immigration' events, the equivalent of changing the immigration rate or altering the island's isolation from the mainland pool. Robinson and Dickerson (1987) found that such changes could affect the species richness, species composition and dominance (relative abundance) patterns in the resulting community. The colonization success or failure of a species could not be predicted merely by its position in the sequence of immigration, although in some cases priority effects were important. The immigration sequence had more influence on relative abundance patterns when immigration rates were relatively low (as on islands) than when they were high (as on continents) and arrival of a species at a time when resource levels were favourable seemed to be more important than early arrival per se. 'The early bird will only get the worm if there is a worm to be had, and if the bird eats fast' (Robinson and Dickerson 1987: 593). They found that from the same species pool substantially different end-point communities could be formed, but the number of such alternative communities was quite restricted. There appeared to be deterministic constraints on the effects of random changes in the immigration schedule, and these were thought to be related to such factors as interspecific competition and predation.

Most of the species that they added persisted for a short time only, but over all their experiments there was a group of early-establishing species, and membership of this early-establishing 'core' changed very little. In some of the sequences that they devised the resulting community became closed, resistant to any further colonization. The existence of such closed communities had been suggested earlier by David Lack as a result of his studies on the avifaunas of Caribbean islands (Lack 1976). The early-establishing core groups may result from the dispersal attributes of the species concerned, as suggested by Gilpin and Diamond (1982) and, for intertidal communities in Australia, by Underwood *et al.* (1983). They may also result from the restricted number of available habitats or resources limiting the number of available niches, as Lack suggested. Lack argued that since the number of available niches is fixed, competitive exclusion would result in assemblages that, in ecological time, are stable and resistant to further invasion, species turnover being restricted to transients that failed to penetrate the now closed community.

Drake (1991) assembled freshwater laboratory communities using nitrifying bacteria, algae, protozoans and crustaceans in his species pool. He found that there were two main types of development 'trajectories' (courses of community development). On the one hand, indeterministic trajectories, which were not repeatable and resulted in alternative community compositions in the replicates, were the result of small differences in the growth rate and clutch sizes of his founder populations. On the other hand, deterministic trajectories were relatively unaffected by variation in founder characteristics and the resulting communities had the same composition of producer species, the same dominance patterns, and the same susceptibility or resistance to the same set of consumer species.

Experimental islands in nature

In an important but frequently overlooked review of the colonization of freshwater ponds, Jack Talling (1951) noted the strong 'chance element' in the initial constitution of pond communities. At that time, he rightly emphasized that most relevant information was based on incidental observations, but Talling also quoted the earlier sentiment of Fritsch (1931), whose discussion of algal communities included the general comment that 'Ponds are at least largely of the nature of isolated islands ... so that a considerable element of chance becomes a factor in determining their flora.' Serious experimental attempts to address this theme started a decade or so later.

In 1963, one of the first attempts to use an experimental approach to the origin and development of island communities was made by Maguire. He created small artificial mini-ponds, aquatic islands, simply by placing bottles of sterile, reconstituted, fresh water at various distances from a potential source biota in a freshwater pond. The bottles of water acted as pristine mini-lakes, and their size and isolation from the potential source could be controlled, and the rate of their colonization by aquatic microorganisms monitored. He found that as the number of species present in the bottles increased (to a maximum of something like 200 species), the rate of colonization declined; the number of potential 'new' bottle-colonists remaining in the source became fewer and fewer. More importantly, Maguire (1963) discovered that the number of different, stable, interacting assemblages of species that developed in his array of aquatic 'islands' was quite limited, although of course the number of theoretical combinations was very great. Certain species were significantly positively, and others significantly negatively, associated. In other words, the presence of certain species may enhance or impede colonization by other species, so that the sequence of arrival may determine later community composition.

Ruth Patrick's experimental project in North America was the next major experimental approach to the problem (Patrick 1967, 1968). She suspended sterile glass laboratory microscope slides in streams, and followed the build-up of diatoms on them over a couple of weeks. She was able to manipulate the area of the 'island' (by altering the area of glass surface available for colonization) and the colonization rate (by varying the current speed). Over the colonizing period, the resultant community contained more species on the larger slides than on the smaller ones, other things being equal, demonstrating the importance of area in determining the richness of a colonizing community. She also found that some colonizing species became extinct from the glass surfaces, while others persisted, and that crowding and competition reduced the number of species in the community as time progressed. Usually it was the rare species that became extinct. (Incidentally, studies such as these work very well as student practical exercises.)

Experiments on the same general principle but using settling plates of various materials immersed in the sea were first carried out in connection with investigations related to the 'fouling' by marine fauna and flora of surfaces on piers and the hulls of vessels (the efficiency of the vessel being thereby greatly reduced by drag). Later such experiments, and others using fragments of plastic sponge as substrates, were used in colonization studies. Osman's (1982) experiments on colonization of fouling panels showed convincingly that rate of colonization, and complexity of the communities which developed, depended on proximity to rich source communities.

However, perhaps the most famous ecological experimental work in connection with colonization studies involved the natural colonization of manipulated but natural islands. Before considering these, however, a digression is needed on the theory that the experiments were designed to test.

The equilibrium theory of island biogeography

In the1960s MacArthur and Wilson put forward an important theoretical concept, the equilibrium theory of island biogeography (ETIB) (MacArthur and Wilson 1967). This theory has dominated considerations of island biogeography for the past four decades. Indeed, in his recent book on the subject, Whittaker (1998), an iconoclastic critic of the theory's usefulness, nevertheless devoted almost a third of the book's text to the ETIB and its implications for island ecology theory and practice. Such has been and still is the influence of this theory.

In this book the scope is narrower than that of the theory. We are concerned with the beginnings of island ecosystems and their early development, and it therefore follows that the examples treated are all pre-equilibrial island systems. Nevertheless, the ETIB's usefulness is not confined, as some would have it, to systems that are at equilibrium. Rosenzweig (1995) showed how the ETIB's dynamic view of diversity helps to explain many cases that are non-equilibrial. Thus before examining individual cases of the assembly of island biotas it is necessary to consider the ETIB and assess its value and relevance.

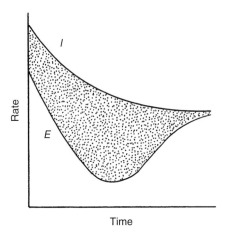

Figure 2.2 Model postulating that, if succession is important in early stages of colonization, the time curve for extinction rate (E) can initially show declines and then rise to the equilibrium level. The immigration rate (I) gradually slows after pioneer stages, but early pioneer communities are not wholly extirpated. (Redrawn from MacArthur and Wilson 1967.)

MacArthur and Wilson argued that as a pristine island was colonized from a given source, although the number of species on the island would rise, the rate of immigration (the number of species new to the island plotted against number of species present on the island) would gradually decline, as fewer and fewer species remain in the source pool as potential new immigrants to the island. When no potential new immigrants remain in the pool, all its species will be present on the island and immigration rate to the island will become zero. The relationship will be expressed graphically as a straight line declining to zero.

Extinction rate from the island (plotting number of extinctions against number of species present) will be zero of course when there are no species present. But as more and more species accrue and are thus available to become extinct, extinction rate will rise. The extinction rate plot will be a straight line rising from zero.

These relationships will result although it is assumed that all species have the same immigration and extinction rates and there is no interaction between species (that is, the presence of a species does not affect the likelihood, one way or another, of any other species being present). Even in this case the island's diversity (number of species) will reach an equilibrium when island immigration and extinction rates are equal. Moreover, this equilibrium will be stable; that is, it will be self-correcting. Species additional to the equilibrium number will result in extinction rate becoming greater than immigration rate and so the number will fall, whereas if the number drops below equilibrium, immigration rate will exceed extinction rate and the number will rise. Species number will be pushed back to the equilibrium whenever it moves above or below it.

MacArthur and Wilson then reasoned that since species are not all identical, the model needed to be modified to take account of the biological properties of species (Fig. 2.2). Species differ, among other things, in colonizing ability. They

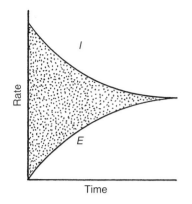
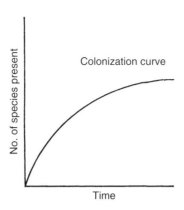

Figure 2.3 Interaction of immigration and extinction: (left) the twin curves for immigration (*I*) and extinction (*E*) over time produce equlibria which (right) produce the form of colonization curve shown (cf. Fig. 2.2). (Redrawn from MacArthur and Wilson 1967.)

assumed that those species with inherently good colonizing ability would be the first to colonize, leaving the potential new colonists remaining in the source pool progressively less and less likely to colonize, and therefore taking progressively longer to do so. So the decline in immigration rate would be greater with the arrival of the early, pioneer species than with the arrival of an equal number of late-comers, and the declining immigration rate curve will then not be a straight line, but will be concave upwards (Fig. 2.3). Second, there will be some interaction between species. The presence of a given species on the island may well affect the likelihood of the presence of another one. Interactions (for example predation and competition) between species would mean that newly arriving species would find it more and more difficult to become established as the number of species increases. Moreover, as the island fills up, ecological saturation will be approached and the island will become 'closed' to some species. This may be because of the upper limit to the number of resource-limited species that its habitats will hold (competitors too similar), the smaller population numbers that follow from an increase in the number of species numbers leading to species becoming too rare to survive, or reduction in predation on the island altering competitive relationships between prey species (or an increase simply preventing an immigrant prey species from reaching a stable population). The likelihood of these factors working to prevent colonization will increase as the number of species on the island rises. For these additional reasons, then, the slope of the declining immigration curve should become progressively less steep as number of species increases, and the curve will be concave upwards.

Again because of the differing biological properties of species, extinction rate should increase at an accelerating rate with number of species. As the number of species competing for the available, finite, island resources (food, space) increases, populations of at least some of them will contract, in some cases to below a viable population size, and when an outside disturbance, such as a volcanic eruption, hurricane or disease, hits their population, these species will become extinct. So extinction rate will accelerate as number of species on the island grows, and the extinction rate curve will also be concave upwards.

As noted above, the theory concludes that where the curves cross (when the rising rate of extinction meets the declining rate of immigration), any increase in immigration rate will be met by an increase in extinction rate, and a slackening of the former will be counteracted by a lowering of the latter. Although the *composition* of the island's biota will continue to change, the *number of species* present will fluctuate between quite narrow limits. Wilson (1992) used the felicitous phrase, 'like an air terminal', to illustrate this dynamic concept, with people (species) constantly arriving and leaving (turnover) but the number in the terminal staying about the same. Note that the theory does not imply a strict one-to-one replacement of extinct species by new immigrants, any more than the departure of a plane-load of passengers implies that an exactly equal number will arrive at the terminal to replace them. An island's equilibrium number of species will be characteristic for a given group of species, and will depend on the island's area and isolation.

MacArthur and Wilson had noted that there was a simple relationship between the areas of islands around the world and the number of species they carry. They saw that larger islands had more species, and the relationship was a simple arithmetic one. As a rule of thumb, number of species doubles with every tenfold increase in area, or the number of species is halved for a tenfold *decrease* in area. This species–area relationship is described by the simple equation, $S = CA^z$, where S is number of species, A area of island, and C a constant. The value of z depends on the group of organisms concerned, and the isolation from source areas of the islands under study; it is constant for a given group of islands and a given group of organisms. It ranges from about 0.15 to 0.35 over all situations that have been analysed, and for the rule of thumb mentioned above, z would be 0.30.

MacArthur and Wilson saw area of island as predominantly affecting extinction rate. Other things being equal, on large islands newly established colonists will be able to form larger, and therefore putatively more secure, populations than they will on small islands. Thus large islands will have the lower extinction rate. Immigration rate will be the same for all (because they are equally isolated), so large islands will have the higher equilibrium number of species. Area will have an effect per se, as just described, but larger islands also tend to have more habitats, an additional reason for large islands to contain more species than small ones.

They saw isolation as affecting largely immigration rate. Again, other things being equal, islands close to a given source will have a higher immigration rate than more distant ones of the same size, which will fill up more slowly (species with poor dispersal powers will take longer to reach them). Extinction rate will be the same on all (because they are the same size), but on distant islands it will

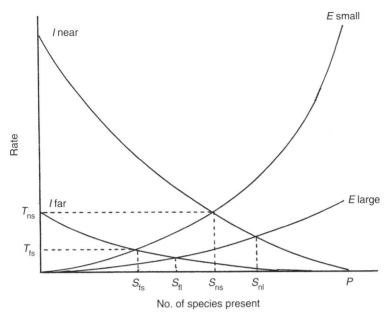

Figure 2.4 A form of the equilibrium theory of island biogeography, after MacArthur and Wilson (1967) and as shown by Whittaker (1998). The diagram shows how immigration rates (I) are predicted to vary as a function of distance, and extinction rates (E) as a function of island area. Each combination of island area and distance (a measure of isolation) produces a unique combination of species number (S) and turnover rate (T). This model demonstrates two cases of how T may vary with different area and isolation. Thus, for a near small island, species number at equilibrium will be S_{ns} and rate of turnover at equilibrium is T_{ns}. P, number of species in pool; fl, far large; fs, far small; nl, near large; ns, near small.

meet immigration rate when fewer species are present. So distant islands will have a lower equilibrium number of species than nearby ones.

Thus, in theory, large islands should have the higher rate of immigration and small ones the higher rate of extinction; and near islands the higher rate of immigration and isolated ones the higher rate of extinction (Fig. 2.4). Small isolated islands will have the greatest extinction rates and smallest equilibrium numbers of species, large near ones the highest immigration rates and highest equilibrium numbers of species.

Note that the equilibrium number is characteristic for a *particular group of species*. In groups of organisms that are both good dispersers and short-lived (allowing extinction to occur rapidly), such as many birds or flying insects, immigration and extinction curves will both be steep, and equilibrium should be reached relatively quickly. At the other extreme of a continuum, organisms with limited dispersal powers and that are long-lived (thus persisting for longer once they become established), such as many large mammals or many trees of mature forest, will reach equilibrium slowly, perhaps permitting human disturbance, climatic change or even geological change to exert a significant effect on colonization. Indeed, the theory's proponents tend to be island zoo-geographers who work with birds or insects, and its strongest critics are island plant geographers.

Since the theory was published there have been a number of refinements. As envisaged by MacArthur and Wilson (1963), island area may affect immigration rate, as well as extinction rate. They did not build this into their mathematical model because they assumed that the effect would be significant only between islands near the source. An island with a large area has a longer coastline and greater surface area for the landfall of any water-borne or aerial propagules, respectively (the effect noted by MacArthur and Wilson). And large islands, as noted above, are usually more diverse, both topographically and ecologically, than small ones, so providing a wider variety of habitats, and conducive to the establishment of a broader array of colonists.

Moreover, isolation can affect extinction rate as well as immigration rate, through an effect now known as the 'rescue effect' from Brown and Kodric-Brown's (1977) work on individual thistle plants as ecological islands for insects. It was earlier noted, as applying to mainland areas, by Preston (1962). Declining populations that are becoming dangerously low in numbers will have a greater chance of replenishment by immigration of conspecifics, and thus escaping extinction, if they are on an island close to the source than if they are on a distant one. Both this refinement and the one in the previous paragraph will tend to push immigration and extinction rates in the same directions as in the basic theory.

Rosenzweig (1995) regarded cases of the rescue effect as violating his bio-logical definition of 'island'. He stipulated that when the population of a species is maintained by immigration of individuals – a 'sink' species (Shmida and Ellner 1984) – the region is not 'self-contained' (Rosenzweig 1995: 211) and is not an island as far as that species is concerned. 'Self-contained' means that species must have average net reproductive rates sufficient to maintain a pop-ulation. Rosenzweig's definition of an island also requires that its species orig-inate entirely by immigration from outside the region. He defined a mainland as a region whose species originate entirely by speciation within it, and regarded the two definitions as extremes of a continuum. So Hawaii, for example, which has many species originating by speciation within the region and yet has (per-haps fewer) species originating by immigration, would lie towards the mainland side of the continuum. Rosenzweig's definition of a mainland also requires it to be, like an island, a 'self-contained region', yet he regarded Dane County, Wisconsin, as a mainland, despite acknowledging that many of its species will be maintained by immigration from neighbouring areas (as Preston pointed out). Rosenzweig's book is an excellent contribution to biodiversity theory and to island biogeography, and is packed with interest and ideas. But I must respectfully disagree with these definitions. They are interesting but, I think, unhelpful. For example, both extremes of the continuum must have no sink species. Then where on the continuum is a region that does have sink species?

MacArthur and Wilson envisaged that competition and predation would be involved in determining the numbers of species on islands, and this has been shown to be the case. Rosenzweig reviewed the evidence in support of their roles. Some authors (but not Rosenzweig) have hailed these findings as some sort of new insight that frees us from the 'species numbers games' (a chapter heading for a treatment of the 'core ETIB' in Whittaker 1998). Several so-called non-equilibrial alternative 'models' have been suggested. They include 'disturbance island theory', 'dispersal-structured model of island recolonization', and 'successional model of island assembly' (*sic*) (Whittaker 1998). These 'models' are claimed to hold more promise and possibly more practical value than the 'games', because of their supposed monopoly of the ability to focus on 'species compositional structure'. I beg to differ on this. They are not alternatives. They are part of the 'game'. Disturbance, differential dispersal, species interactions and succession were all treated by MacArthur and Wilson. The last three were integral parts of their model.

'Disturbance island theory' centres on a generalized graph of pre-equilibrial Krakatau plant colonization data which is then speculatively extrapolated to show what may (or may not) happen if a future intermediate- to high-magnitude disturbance event, such as a hurricane, should strike the islands. Bush and Whittaker (1993) believed that such an event might cast the system away from equilibrium, such that it is 'permanently non-equilibrial'. Ward and I, however, pointed out that in 1988 a very destructive hurricane did not result in any species extinctions in the Nicaraguan primary rainforest (Zimmerman *et al.* 1996), and that in Florida in 1995, 5 months after one of the most severe hurricanes of the century (Andrew), there was no effect on the seemingly highly susceptible and fragile obligate mutualist associations between fig species and their fig-wasp pollinators (Bronstein and Hossaert-McKey 1995, Ward and Thornton 1998). We concluded that the only event likely to have the effect envisaged by Bush and Whittaker is a cataclysm of the scale and intensity of the Krakatau 1883 eruption. And in such a case the system would return to zero and the whole process of assembly would start again.

The 'dispersal-structured model of island recolonization' is illustrated by a group of diagrams showing the plant dispersal-mode spectra for various stages of the floral reassembly process on Krakatau, and is unlikely to apply, even broadly, to islands in general. The 'successional model of island assembly' is basically a statement that developing biotas proceed through a series of successional phases. The ETIB does not assume otherwise.

The alternative 'models' are in fact simply factors involved in the assembly process of the ETIB model, several of which were explicitly noted by its authors.

Thus, the differing dispersal powers of potential immigrant species and inter-actions between species were important reasons for the concave shapes of the theory's immigration and extinction curves, respectively. Also, MacArthur and Wilson (1967: my Fig. 2.2) illustrated the effects of succession by showing that during its early stages, when rate of immigration outstrips rate of extinction, the extinction curve may not be rising monotonically, but may at first decline. They suggested that this may also be the case where there is a delay in the inhibitory effects of some species on others, such as the time taken for forest trees to shade out pioneers, or where succession (of birds, for example) involves different trophic levels.

Although the 'alternative' models are claimed to be 'much more advanced than they were in the 1960s', it is difficult to regard any of them as rigorous, testable, mathematical models comparable in scope and general applicability to the ETIB. In my view, as presently formulated, they form no theoretical alter-native to it, even collectively.

Recent criticisms of the ETIB included the following statements: 'in the "narrow sense", the ETIB has been refuted enough times for it to be dead and buried' and 'The mantra that only equilibrium models will bring salvation should finally be discarded' (Whittaker 1998: 143, 187). Rosenzweig (1995) did not agree with the first statement. After extensively reviewing, and in some cases reworking, various tests of the theory, he concluded (p. 263): 'The theory of island biogeography holds up well.' The second statement is rather like the 'Have you stopped beating your wife?' conundrum. I know of no biogeographer who holds the specified 'mantra'. Within the statement is an unjustified assumption, a straw man, which is then demolished. The statement is therefore meaningless.

But it is necessary briefly to examine some serious criticisms of the ETIB. They are:

(1) *The ETIB treats all species as being interchangeable, non-interacting units.*
     It has been fashionable to criticize the theory as being non-biological, treating species as simply numbers, without regard to the ecological pro-cesses involved in colonization, and such criticisms are still to be found. Were species really treated as inanimate interchangeable items, without interactions, both immigration and extinction curves would have been straight lines (and the theory would still have applied). As we have seen, however, the hollow immigration and extinction rate curves are based on assumptions of biological attributes and differences between species.
(2) *The ETIB ignores successional effects.*
     MacArthur and Wilson (1967) recognized (p. 21) that more rapidly dispers-ing species were likely to become established before more slowly dispers-ing ones (hence the immigration rate curve is a falling concave one). They

also recognized (p. 51, figure 23) that 'when plants as a whole are colonising a barren area, the extinction [rate] curve should decline at first due to succession', because facilitation is likely to take effect before the inhibition produced by plants shading out others. They also acknowledged (p. 51) that where there is a succession involving different trophic levels, the time curve for extinction rate could also initially decline.

MacArthur and Wilson were also well aware that in the lead-up to equilibrium most of the turnover would be successional, and (p. 47) cited the loss (since confirmed to have been true extinctions) from Krakatau in the 1920s and 1930s of several birds, losses for which 'change in habitat as the island became forested' was thought to be partly responsible.

(3) *Most turnover is either successional or involves transients; it is not the 'stochastic' turnover that the theory demands.*

That species are replaced during succession was explicitly acknowledged by the authors of the ETIB (item 2 above). Certainly, in many cases turnover appears to be heterogeneous (the likelihood of extinction not being the same for all species), with transients making up a good proportion of it. This does not destroy the theory, nor, as has also been claimed, render it trivial. The ETIB acknowledged that species interactions would be likely to result in population declines of at least some species, thereby increasing their likelihood of extinction, and that this would have a greater effect on small islands, for example. It is neither surprising nor destructive to the theory that the first species to become extinct are likely to be those that are the least well established, without secure populations, or those whose island populations are normally small.

(4) *Turnover at equilibrium has not been demonstrated.*

Some 'failed' attempts to demonstrate equilibrial turnover have involved pre-equilibrial situations involving slowly reproducing, long-lived organisms (e.g. Krakatau forest trees). In other cases, where turnover rate at equilibrium showed no strong relationship to either area or isolation, as predicted by the ETIB, the organisms studied almost certainly have dispersal powers, as a group, very much greater than the distance between source and island (for example, orb-weaving spiders on very small central islands of the Bahamas: Toft and Schoener 1983). It is true, though, that it has been difficult to demonstrate turnover in equilibrial situations, probably because studies of non-experimental equilibrium situations have been insufficiently long-term. The Florida Keys experiments of Simberloff and Wilson (p. 25) remain the best demonstration of turnover at equilibrium. This experiment involved organisms with a variety of dispersal powers and fast life cycles. As expected under the ETIB, turnover rates were initially high, and declined as the assortative phase of biotic assembly was completed.

(5) *The ETIB ignores historical data and the role of environmental change.*
     No general model could include historical factors, which by definition are
     peculiar to each individual case. Environmental change through succession
     is treated (see item 2 above). As Rosenzweig (1995: 246) put it, 'The model
     contains some variables, but not all real variables. Thus we try to isolate and
     assess the influence of the variables dealt with in the model. Such tactics
     may not always work, but we do not make simplifying assumptions to
     claim that there are no other variables.' The fact that the ETIB does not
     explicitly incorporate these largely idiosyncratic factors is not a weakness.
     In fact, MacArthur and Wilson (1967: 64) cited one of the virtues of the ETIB
     as 'making the individual vagaries of island history seem somewhat less
     important in understanding the diversity of the island's species', although
     they acknowledged that 'Of course the history of islands remains crucial to
     the understanding of the taxonomic composition of species.'

(6) *The ETIB ignores 'hierarchical links between taxa'.*
     If I understand the (mis)usage of this phrase correctly, it refers to the well-
     known ecological facts that some taxa feed on others, some displace others,
     some pollinate others, and so on. It hardly seems necessary to state that the
     authors of the ETIB were well aware of these relationships, and that they
     affect immigration and extinction rates. Chapters 4 and 5 of the MacArthur
     and Wilson book (on 'The strategy of colonization' and 'Invasibility and the
     variable niche') provide many examples of the way in which such links may
     affect colonization. These effects were not included explicitly in the ETIB for
     the same reason that they have not been included in any so-called 'alterna-
     tive model'; they are difficult to quantify in the context of the model. Some
     were, however, included in Diamond's assembly rules (see below, Chapter 7).

(7) *Non-monotonic curves, such as some found in the Krakatau data, contradict the ETIB*
     because they imply the existence of alternative stable equilibria.
     They do not necessarily imply this. Ward and I showed (1998) that non-
     monotonicity does not inevitably produce alternative equilibria. And, even
     if alternative equilibria do occur, the first one reached is likely to be either
     so transient as to be undetectable or so stable as to be effectively perma-
     nent. So alternative equilibria would pose only a minor threat, at most,
     to the ETIB. The possibility of non-monotonic curves was envisaged by
     MacArthur and Wilson.

(8) *The ETIB ignores evolution.*
     Chapter 7 of the MacArthur and Wilson book, 'Evolutionary changes fol-
     lowing colonization', takes up 35 of the 177 pages of text, a fifth of the
     book. The quantitative effects of evolution were not included in the basic
     theory, which pertained to an ecological timescale. In a later publication,
     MacArthur (1970) provided a sketch showing the relationship of evolu-
     tionary processes to the ecological ones.

The mangrove islet experiments

Wilson considered how to test the ETIB. Inspired by Krakatau's recolonization after its devastating 1883 eruption, with Daniel Simberloff he conceived the idea of creating a number of 'miniature Krakataus' so that recolonization could be studied while important parameters of the recolonization process were experimentally manipulated. In the late 1960s and the 1970s Simberloff and Wilson carried out what was to become the most famous field experiment on island recolonization. Their experimental 'mini-Krakataus' were small patches of red mangroves (*Rhizophora mangle*) growing in mud in the Florida Keys (Simberloff and Wilson 1969, 1970, Simberloff 1976). The mangroves were substrates for a considerable invertebrate animal community. Simberloff and Wilson artificially deprived some of the mangrove islets of their fauna by enclosing them in plastic tents and fumigating them with methyl bromide. They then followed the recolonization of these 'defaunated' islets by monitoring them several times during the first year, and again after 2 and 3 years.

Although the mangrove islets were characterized as 'miniature Krakataus', there is an important difference. The islets were defaunated but were not also 'deflorated'. The mangrove plants themselves remained, alive. Indeed, the plants formed the islands, and acted as a living substrate for the recolonizing animals, and as a food source for some.

Islands of various sizes and distances from the main body of mangroves (as the nearest possible source) were used, and their size could be changed during the experiment, for example, by physically cutting an island into two, without altering habitat diversity a great deal (mangrove islets are fairly uniform habitats). The 'defaunated' islands regained their former diversities within a year. Those that were rich before the experiment regained their richness; those that were poor regained their poverty. But the species complements were not exactly reinstated. The new faunas consisted of a different combination of species, including only about a third of the original complement. Like Patrick, Simberloff and Wilson found that the rate of extinction from an islet depended on species density. They showed that competition was involved and that there was differential loss of species. Species numbers on the recovering islands tended at first to overshoot their former equilibria; early arrivers did not become extinct until some time after later arrivers (which were to become successful later in the process) became established. Species numbers fluctuated about an equilibrium until the experiment ended, 2 years after fumigation, and turnover was quite high. These experimental studies made important contributions to the theory of community assembly on islands. To what extent are their findings borne out in completely natural situations?

PART II

Natural recolonization after devastation

CHAPTER THREE

A clean slate?

Volcanic effects on living communities

In this part of the book I attempt to review what is known about the destructive and constructive processes involved in the interaction between volcanoes and living communities, on an ecological rather than an evolutionary timescale. I will pay special attention to those cases on which I have worked, or of which I have been an avid student. I have been greatly assisted by discussions with John Edwards, who kindly sent me the manuscript of his own excellent review on a very similar topic, focusing particularly on his own study area, Mount St Helens (Edwards *et al.* 1986, Edwards 1988, 1996), Eldon Ball, who encouraged me to try to further his own highly significant work on Motmot and Diamond's classic studies on Long Island off New Guinea, and Sturla Fridriksson, who introduced me to Surtsey and has kept me up to date with his team's fundamentally important precise and intensive long-term studies there. In treating the colonization of Surtsey I have drawn freely from Fridriksson's masterly 1975 and 1994 books and other papers detailing the work of his group, largely published in the *Reports* of the Surtsey Research Society.

Obviously, volcanic activity can damage and may even destroy natural living communities. But anyone who has visited an extinct volcano will have seen that sooner or later life returns to devastated areas. Indeed, from a consideration of the biotas of volcanic islands like the Hawaiian group, we know that, given time, rich and complex land communities may develop on substrates that must have been devoid of life, certainly of land life, when they first broke the sea's surface as steaming piles of ash or lava. The constructive power of nature can be seen to be as powerful as, if slower than, its destructive force.

The damage caused to living organisms by volcanic activity clearly depends on the type of eruption. Damage may result from the effects of toxic volcanic gases, explosive blast, debris avalanche, pyroclastic flows, mud flows (lahars), lava flows, or tephra (including ash) fall.

Toxic gases

Perhaps the most well-known case of toxic gases damaging the biota is that of the 7-month-long effusive eruption of Lakagigar in Iceland in 1783, when SO_2, SO_3,

CO_2 and HF emissions destroyed birch woods 200 km away from the vent and were mainly responsible for so stunting the grass crop that about 70% of livestock and 20% of the human population of Iceland perished (Thorarinsson 1971).

Explosive blast

The effects of blast were clearly seen at Mount St Helens in 1980. There were two blasts, a vertical column at the crater itself and a lateral one below this on the north face. The blast that broke through the volcano's northern flank travelled at 1000 km/hr and, with a temperature of 300 °C, devastated an area extending for more than 20 km north of the former summit and for nearly 30 km from east to west. In an area of almost complete destruction, termed the 'blast zone', trees were felled and the vegetation razed to the ground, creating a substrate of bare mineral surfaces on which only primary colonists could become established. This area, extending for almost 10 km from the summit, was the slowest to recover; 5 years after the eruption it appeared much as it did in the immediate aftermath. The 'scorch zone' extended for a radius of some 40 km from the crater in a northern arc. At least 600 km^2 were devastated. All trees, some over 1.3 m in diameter, were snapped off like matchwood in more than 384 km^2 of forest, up to distances more than 32 km from the eruption centre.

Mount St Helens

For the past 7 million years a sub-plate of the great Pacific plate has been slowly thrusting beneath the continental North American plate, creating the Cascade volcanic arc which stretches for 960 km from British Columbia to northern California. Mount St Helens in the state of Washington, USA, is the youngest volcano in this arc. Until 18 May 1980, the mountain was known as the 'Fujiyama of the West' because of its almost perfectly symmetrical cone, and Spirit Lake, beneath its northern flank, was famous for its beauty and its trout fishing. The volcano was regarded as dormant; there had been no significant activity since the Klickitat people watched an eruption in 1857. But in March 1980 seismometers registered earthquakes, and by May the north flank of the mountain was bulging some 150 m under the pressure of the magma beneath; the deformation was apparent to the naked eye. On 18 May, at 08:32 hrs, the top 450 m of the northern side of the volcanic dome collapsed, at first as the slippage of a huge coherent block, but quickly fragmenting into a massive landslide, and 11 seconds later the volcano exploded with the force of 50 Hiroshima bombs.

Mount St Helens is within easy reach of several research centres, and biotic damage and recovery were studied intensively from the time of the eruption (Keller 1986, Swanson 1987). Immediately to the north of the volcano an area of about 50 km^2 at about 1000 m altitude, now known as

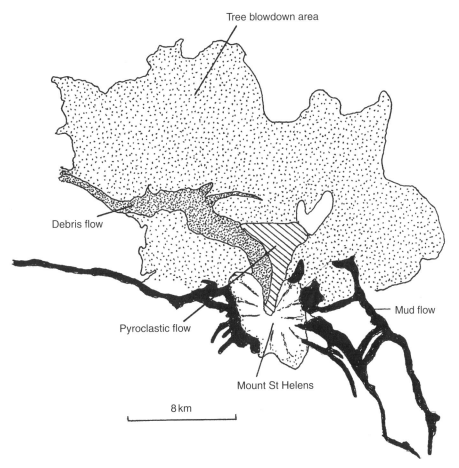

Figure 3.1 The area around Mount St Helens, Washington State, to indicate major impacts of the 1980 eruption. (After del Moral and Bliss 1993.)

the Pumice Plain, was covered in landslide and pyroclastic deposits and its macrobiota was completely destroyed, making it an ideal place to monitor the return of living organisms in the absence of resident populations (Fig. 3.1).

The eruption column extended for 19 km into the atmosphere and half a billion tons of ash fell on an area of 56 000 km^2. Fifty-seven people were killed.

Debris avalanche
Debris avalanches usually contain both large blocks and smaller fragments, and may reach very high speeds. One of the best-documented debris avalanches occurred on Bandai-san, Japan, in 1888. Horizontally directed major explosions

shattered the cone in less than a minute and with a horizontal blast the northern flank of the volcano collapsed in a huge, 1.5-km^3 avalanche which buried several villages and killed 461 people. The erstwhile near-symmetrical crater now had a 700-m-deep gash in its northern rim.

Once again, Mount St Helens provided an infamous example. Within 10 minutes of the start of the eruption, a massive debris avalanche had deposited ash and rocks along a course extending some 27 km westward along the North Fork Toutle river, filling 64 km^2 of the valley with volcanic rubble to an average depth of 50 m. This is one of the longest debris avalanches on record and had a volume of almost 3 km^3.

In 1995 a debris avalanche on Montserrat's volcano covered over one-fifth of the island in debris. During the following year another flow travelling at more than 90 km/hr moved over 2.5 km and extended some distance out to sea.

Pyroclastic flows
Pyroclastic flows (previously known as *nuées ardentes* or glowing clouds) are infamously dangerous. These turbulent, ground-hugging, expanding clouds of fluidized gas-charged incandescent particles move silently and rapidly (up to 100 km/hr) on a cushion of their own continuously exsolved gases. They are lethal to animals and death comes rapidly. They are perhaps the most dangerous to humans of all volcanic effects, as was seen by their fatal impact in 1883 when they swept across the sea surface from Krakatau to the coastal village of Kalianda in Sumatra, 40 km away, killing about a thousand people, or in 1991 on Unzen in Japan, where they killed a group of volcanologists as well as their would-be rescuers. Pyroclastic flows are a feature of Peléan eruptions, which take their name from the eruption in 1902 of Mount Pelée on the West Indian island of Martinique. Here, a pyroclastic flow raced silently down the volcano's slopes to engulf and incinerate the town of St Pierre, killing all but two of the 30 000 inhabitants (one of whom was a prisoner lucky to have been in solitary confinement, incarcerated in a partially underground cell with concrete walls a foot thick).

In the same year the eruption of La Soufrière on the island of St Vincent, 144 km to the south, killed some 1350 people of the (fortunately) sparse population, devastating about a third of its surface, including an extensive area of fertile cultivated land. Most fatalities were caused by mud flows and pyroclastic flows from the base of the eruption column travelling radially as surges down the volcano's flanks. The depth of ash and pumice was from 17 to 27 m in some valleys, from 0.3 to 1.7 m on fairly level ground, and only a few centimetres on steep slopes. One such eruption on Taal volcano in the Philippines in 1911 totally destroyed the vegetation over most of Volcano Island in Taal's crater lake, although the ash deposits were neither very deep (average about 30 cm, drifting to 2 m in places) nor very hot, and several villages and their human and animal inhabitants were annihilated with the loss of over a thousand human lives.

The work of Moore (1967) and Waters and Fisher (1971) on the short, violent eruption of Taal in September 1965 focused studies on base surges, which were a feature of this eruption. The surges are low-density, turbulent clouds rolling away close to the ground at high velocity radially from the base of an over-loaded, collapsing vertical eruption column. They seem to be produced parti-cularly in eruptions involving interaction between magma and water. In 1965 powerful horizontal blasts surged with hurricane force from the base of Taal's column, moving over the surface of Lake Taal to its shores. Trees less than a kilometre from the vent were snapped off or uprooted; those further away were defoliated and the ash-laden blasts scoured up to 10 cm of wood from the trunks on the side facing the vent. The trees were not burnt or charred, however. The violent surges could not have been much hotter than 100 °C, for there were thick deposits of mud at the edge of the devastated area, about 6 km from the vent. Deposits accumulated to a depth of 4 m within 2 km of the vent, and up to 7 km away objects were sandblasted by the lower component of the base surges.

The temperature of pyroclastic flow clouds can reach 600–700 °C but the heat is experienced for only a few moments; victims in St Pierre suffered extensive body burns although their clothing was not even singed. On Mayon volcano, in the Philippines, pyroclastic flows in 1968 were surrounded by a 'seared zone' from a few metres to more than 2 km wide. All animals within this zone were killed outright, presumably from burns or by inhaling the fluidized incandes-cent cloud. On the inner parts of the seared zone all vegetation was charred and many trees stripped of foliage on the side facing the crater. Plant leaves were shrivelled and browned at its outer edges. At Mount St Helens, pyroclastic flows of ash and superheated gas moved swiftly down the flanks, north to Spirit Lake some 8 km from the crater and westward for a similar distance to a bend in the North Fork Toutle river.

Lahars

Mud flows (lahars; lahar is an Indonesian word) can also be devastating. When Bali's Gunung Agung erupted in 1963, 1184 people died in the ash flows and mud flows. At Kelut volcano on Java, lahars have caused many deaths and have led to large-scale tunnelling works on the mountain to alter lake levels and lessen the likelihood of dangerous flows. Cotapaxi in Ecuador is another 'water volcano'. In its 1742 eruption the ice on its summit was melted and mud flows engulfed 800 people. It was estimated that the waves of the flow were over 20 m high and travelled at more than 45 km/hr. During another eruption in 1877, one of the first pyroclastic flows to be recorded melted the ice, causing gigantic mud flows. Perhaps the most damaging mud flows on record were those of Nevado del Ruiz, in Colombia in 1985. Here, explosions caused the summit glacier to melt and the resulting enormous mud flows engulfed the town of Armero and its 22 000 inhabitants. Infuriatingly, the eruption had been predicted but volcanologists could not convince the authorities of the imminent danger. At Mount St Helens, mud flows radiated from the crater to the south as far as 16 km and to the west for 30 km along the South Fork Toutle river.

Mud flows are also sometimes formed when dust clouds rising from pyroclastic flows trigger torrential rainstorms within minutes of the passage of a *nuée*, as on Mayon in 1968.

Lava flows

The effect of lava is 'all or nothing'. Clearly lava flowing over an area will obliterate and kill all surface organisms that are unable to move out of its path. Although trees only a metre from the edge of a recent flow on the active emergent volcano Anak Krakatau survived without any evident ill effects, nothing on the surface covered by a flow is known to have survived.

Ash (tephra) fall

In reviewing the effect of damage caused by tephra fall in some of the large Icelandic eruptions, Thorarinsson (1971) concluded that the damaging effect depends particularly on the thickness of the fall, its timing, and its chemical and physical properties. He found that 10 cm of freshly fallen tephra resulted in abandonment of farms for a year, ash 15–20 cm deep led to abandonment for from 1 to 5 years, and where the thickness was 20–50 cm settlements were abandoned for decades. He cited the 1902 eruption of Santa Maria in Guatemala, when coffee plants in areas that received less than a metre of tephra recovered within a year, although grazing areas were seriously damaged where tephra was over 15 cm thick. In the 1912 Katmai eruption in Alaska, in an area covered by 50 cm or more of tephra extending for 60 km to windward of the volcano, most trees were killed, and at Kodiak, 160 km from the volcano, a 30-cm-thick layer caused extreme damage to crops. Susceptibility to damage by tephra and gases varies widely between plant species.

Tephra can be removed by erosion, of course, and rain erosion in particular can be quite rapid on volcanic slopes. After the 1964 eruption of Irazu volcano, Costa Rica, from a half to a third of newly fallen tephra was lost from the upper slopes within a year (Waldron 1967) and more than half the tephra deposited around La Soufrière in 1902 had been eroded to the ocean within 6 months.

Shortly after the Paricutin eruptions of 1943–1952 in Mexico, ash samples were invariably found to have at least minimal amounts of nitrogen to nourish plants, and a significant part of it was in the form of soluble compounds. In areas impacted upon by tephra, vertebrates were lost (Burt 1961), but by 1959 erosion had freed the former forest floors of ash, except where it was protected by a blanket of conifer needles (Eggler 1948, 1959, 1963). Where the ash had been eroded away, vegetation reappeared as pre-existing plants resumed growth and new plant colonists were established. On the cinder cone, vascular plants were first seen (two species) in 1957, and by 1960 there were 28 species.

In a review of the effects of the 1980 Mount St Helens eruption, Edwards (1996) noted that living communities in subalpine meadows covered by up to 20 cm of tephra in spring appeared to have returned to normal by the end of the summer. Regrowth penetrated the ash layer during summer, and terrestrial vertebrates, birds and insects recolonized with the returning plant cover. In Alaska small mammals such as squirrels, mice and marmots were killed by tephra falls from 2 to 10 cm deep. Tephra deposits that are several metres thick, however, can maintain temperatures of from 600 to 800 °C for weeks and months, and it is thought that survival chances of the macrobiota under such conditions are practically zero.

The developing communities recently present on the three coastal ash fore-lands of the emergent island Anak Krakatau (which emerged from the sea in 1930 and suffered self-sterilizing eruptions in 1952–53) are not in seral syn-chrony. The developmental stage of the north foreland, which from 1979 to 1989 was largely covered in grassland, in 1992 was rapidly changing to *Casuarina* woodland, was about 15 years behind that of the east foreland, where *Casuarina* woodland was well developed in 1992 and beginning the transition to mixed forest; the north-east foreland was intermediate between the two (Thornton and Walsh 1992). This successional mosaic appears to result from differentially damaging eruptions in 1972–73 and involves not only plants but also birds, spiders, butterflies, tropical fruit flies and invertebrates feeding on microepi-phytes on the bark and needles of *Casuarina* trees. Whilst volcanic fumes may have affected the forelands differentially, most of the differences are probably the result of differential exposure to ash fall.

Tephra fallout can have a significant and special effect on insects. The Mount St Helens tephra, for instance, consisted of a fine powder of volcanic glass fragments which acted as a natural insecticide, abrading the insects' outer lipid layer of epicuticle and causing fatal dehydration. In the laboratory Edwards and Schwarz (1981) compared the effects of this tephra and smooth-grained river sand on a number of insect species. Insects exposed to tephra, but not those exposed to river sand, lost water and died within 30 hr. On the volcano, Colorado potato beetles (*Leptinotarsa decemlineata*) died within 24 hr of exposure to tephra unless they happened to have been washed clean by over-head irrigation systems, and an economically significant outbreak of grasshop-pers was controlled by the fallout, obviating the need for aerial spraying with insecticide (Edwards 1996). In contrast to the effects of Mount St Helens tephra, however, in September 1996 daily fallout of Anak Krakatau's coarser-grained basaltic ash on Sertung, an island 3–4 km downwind from the vent, had no obvious effect on flight activity of grasshoppers, carpenter bees or dragonflies.

These insects may have entered the ashfall area from elsewhere; on Mount St Helens vespid wasps and flies were flying over barren surfaces the day after the eruption.

Edwards and his colleagues investigated the effects of Anak Krakatau's tephra on a test animal, the house cricket, *Acheta domestica*, in the laboratory (Edwards 2005). As expected, exposure increased respiration rate and water loss, and decreased longevity, but the effects were less marked than when the test insects were exposed to ash from the Mount St Helens 1980 eruption. The latter eruption caused widespread insect mortality in areas of ash fall (Edwards 2005). Physical texture of the particles is of prime importance for ash to have insecticidal effects (Edwards and Schwartz 1981), and the dark grey basaltic Anak Krakatau tephra is more granular than the near-white, silica-rich powdery dacitic tephra of Mount St Helens. It is not clear how this difference relates to the observed difference in effects. The finer Mount St Helens tephra may penetrate more deeply into the insect's tracheal system, or may interfere with the waterproofing mechanism of the cuticle not only through a more damaging abrasion of the cuticular waxes but also by absorbing them or altering their structure. Although exposure to both Mount St Helens ash and ash from Anak Krakatau had a detrimental effect on survival in the laboratory, the significant difference between the effects of ash from the two volcanoes on the survival of test insects suggests that caution must be exercised in making general statements about the effects of ash fall on insects. In Java, some caterpillars were unable to feed on vegetation covered with ash from local volcanic eruptions, and died (Gennardus 1983). In summary, insect mortality from ash has a number of causes. Pyle (1984) noted entrapment, abrasion of cuticle and body surface waxes causing water loss, desiccation enhanced by absorptive properties of ash, and excessive salivation during grooming. Ingestion of ash by caterpillars can slow their rate of development and reduce survival rates, as well as influencing their parasites (Bromenshenk *et al.* 1987).

Apart from the effects of heat, burial and the physical effects on insect cuticle, the chemical nature of volcanic fallout may be an important determinant of its lethality. For example, acidic volcanic glass is more stable when exposed to chemical weathering and poorer in nutrient and trace elements than is more basic glass. Sometimes chemical poisoning occurs. Fluorine, for example, becomes concentrated on the surface of ash grains, and fluorine poisoning of livestock is a fairly common hazard in Iceland. In the 1970 Hekla eruption, ash containing high concentrations of leachable fluorine fell over an area of $20\,000\,\text{km}^2$ in the first two hours, poisoning more than 100 000 sheep and cattle and being largely responsible for the toll in human lives (Thorarinsson 1971). A layer of ash hardly 1 mm in thickness was sufficient to kill grazing sheep 170 km from the eruption, and Thorarinsson noted that in a drier climate the fluorine would have persisted on the surface for longer.

Grain size affects the damage directly caused by tephra, and is also important, for example, in relation to fluorine poisoning. Whereas coarse-grained wind-blown tephra, such as that of the 1875 Askja eruption in Iceland, can tear vegetation and destroy grazing, finer-grained ash deposited further from the volcano may carry concentrations of fluorine (20 000 ppm) 20 times higher than those of the coarser ash falling close to the source (Galanopoulos 1971). At Mount St Helens, in contrast, fodder containing up to 10% by weight of ash was not toxic to livestock, milk production and growth rate of dairy cattle was unaffected, and day-old chicks given tephra as 30% by weight of their food suffered no additional mortality. Over most of the fallout area plants and animals seem to have resumed their normal activities fairly quickly. Clearly, as noted above, generalizations about eruptive effects can be dangerous, and extrapolations from one environment to another are likely to be unreliable.

The timing of the tephra fall may be crucially important. A thin fine-grained layer deposited during the growing season may have a greater effect than a thicker layer falling at other times. The greatest damage to livestock in Iceland occurs when ash falls on grazing areas in spring and summer. As illustrated at Mount St Helens, tephra falling on to a winter covering of snow (or ice) under which the living community is often dormant will have far less damaging effects than would the same fallout in the summer, when the community is active and lacks this protection (Edwards and Banko 1976, Edwards 1996).

Local survival

After the eruption of Krakatau Island in 1883 (Figs. 3.2, 3.3) there was no question of immigration from contiguous unaffected source areas. Nevertheless the question of whether the recovery was from foci of survivors or from overseas colonists was the topic of considerable international debate for some years (Thornton 1996a). The speed and nature of the subsequent recovery of communities on the Krakatau remnant and its two closely neighbouring devastated islands strongly indicated that the macrobiota, at least, had been eradicated and recovery was from new colonists. The question of bacterial survival under deep hot ash deposits is as yet unresolved and must be regarded as a possibility, and some species of soil nematodes are also capable of surviving very high temperatures. Docters van Leeuwen (1936) and Dammerman (1948) successfully argued, however, that the first organisms to appear were not those that would have been likely to survive the eruption by persisting underground, for example, but those with good dispersal powers. In contrast to Krakatau, the first plants to appear on Sebesi, an island about 15 km from the eruption and which received a much shallower deposit of ash than Krakatau, were largely species with underground storage organs or vegetative structures such as rhizomes, bulbs and tubers; they were survivors, not new colonists. Dammerman (1948) pointed out that even if animals had survived on the Krakataus they would be

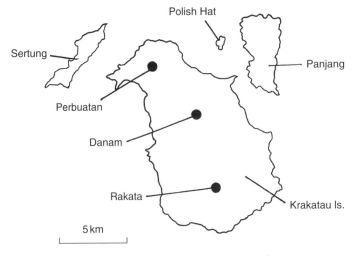

Figure 3.2 The Krakatau archipelago before the eruption of 1883 (cf. Fig. 3.3). The three volcanoes on the main island (Krakatau) are shown; the small island of Polish Hat was obliterated by the eruption.

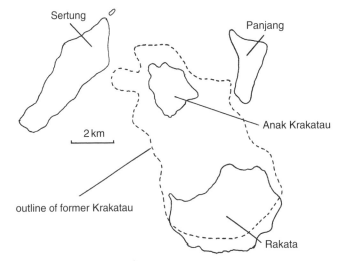

Figure 3.3 The Krakatau archipelago in the 1990s, with outline of former Krakatau indicated. The central island, Anak Krakatau, emerged from the caldera only in the 1930s and is a key focus for study of the themes treated in this book.

confronted with the problem of obtaining sufficient food in a devastated land-scape. This point is perhaps illustrated by two observations on ants, one in Japan, the other in North America, cited by Edwards (1996). Two species of *Formica* were active 9 days after ash fall in the 1929 eruption of Komagatake (Kuwayama 1929), but lasted for only another 11 days; and 6 weeks after the Mount St Helens eruption carpenter ants were taking dead wind-blown insects from the tephra surface on the edge of the blast zone, but were not seen there again (Edwards *et al.* 1986).

Although lava flows and lahars are usually lethal to organisms in their path, a plant or animal may be saved by being sheltered in the lee of natural obstacles, by being on high ground, when the flow will move around it (leaving a *kipuka* – a habitat island of unaffected landscape), or of course, in the case of animals, by moving out of the way of the flow. Although crannies and overhangs may provide shelter from tephra fallout, survival of macroorganisms buried under several metres of hot ash is unlikely. Obstacles may also shelter organisms in their lee from the effects of pyroclastic flows. In 1883 the Ketimbang area in Sumatra was devastated by pyroclastic flows travelling some 40 km across the sea surface from Krakatau, whereas just a kilometre or so away, in the lee of the intervening island of Sebesi, there was no such damage.

As Edwards (1996) pointed out, the Mount St Helens event highlighted the importance of altitude for survival. Snowpack insulates small organisms, protecting them from desiccation and heat, and also moistens and compacts the tephra, rendering it less likely to abrade the cuticular surface of arthropods (Edwards 1986). An ice cover can protect freshwater organisms.

Plants

When pre-existing plants survive an eruption, this is generally more important than the establishment of new immigrants in the recovery of vegetation. The distinction between total eradication and the local survival of a few individuals, although seemingly trivial, is also important to the biologist considering the nature of the process of reconstruction of a natural community. Spread from survival foci is very different from colonization of a *tabula rasa* by immigants from outside the devastated area, and the species assembly process, as well as the nature of the resultant community, may be quite different in the two cases.

In a review of plant survival following burial in tephra, Antos and Zobel (1987) noted that deposits, particularly ash deposits, erode very quickly in heavy rainfall, and on Katmai, Paricutin, La Soufrière, Taal and New Guinean volcanoes this facilitated the recovery of buried plants. There are many reports of herbaceous plants surviving burial under ash to depths of more than 30 cm, and the grass *Saccharum spontaneum* survived burial to 1.5 m in New Guinea. After the Katmai eruption most plant species withstood burial for a year or more; some recovered after 3 years. Herbaceous plants may take up to 4 years to penetrate the deposit and a growth form of 'runners' (such as in *Saccharum*) or adventitious roots is related to survival ability. The growth of woody plants may sometimes actually increase after burial in tephra, in spite of the low concentration of available nutrients. On Hawaii, Paricutin, Katmai and Taal, trees have recovered from a 2-m-thick tephra fall.

On Mount St Helens buried trees whose roots were exposed by rapid rain erosion resprouted, stabilizing the ash slopes. But on this volcano the question of survival was in some cases complicated by the possibility of local reinvasion from contiguous unaffected areas. For example, lupins, which appeared early on the cooled pyroclastic flows, may have grown from viable seed that survived the eruption in sheltered sites or from wind-borne seed dispersed into the area after the eruption (del Moral and Wood 1986, del Moral and Bliss 1987, del Moral *et al.* 1995).

In a recent episode of eruptions of Anak Krakatau, although there was considerable mortality in populations of early pioneer fig tree species, conditions were created that were favourable to the germination and growth of two species (Thornton *et al.* 2000). These were also found to have unusually synchronous fruiting, possibly a consequence of the sudden amelioration of conditions following volcanic activity.

Plants may also survive as seeds. Seed banks in the soil can survive beneath ash deposits, and plants have been grown from seeds found in soil covered for several years by several centimetres of Anak Krakatau ash (Whittaker *et al.* 1992b, 1995). Had such seeds been naturally exposed by erosion of their ash covering, germination would have been possible *in situ*. Whether seeds could remain viable under the very much deeper and therefore more persistently hot layers resulting from cataclysmic eruptions such as those of Thera (Chapter 6), Long Island (Chapter 7) or Krakatau (Chapter 8), is doubtful.

Animals

On Mount St Helens, Edwards (1996) recorded the survival of a grylloblattid and 15 species of non-ballooning spiders sheltering in rock debris at the foot of a cliff in an area where tephra fallout reached a depth of 10 cm. He also cited the results of a transect cut through tephra by R. Sugg in 1982: nine ant species were taken at depths averaging 15 cm, but only one from depths averaging 50 cm (Sugg 1986, Crawford *et al.* 1995). A carabid beetle with reduced eyes and wings and which lives underground was captured at a pyroclastic site in 1983; it may have been a survivor, but Edwards (1996) thought immigration by wind a more likely explanation.

Animals' lifestyles may also protect them. The Mount St Helens eruption had no long-term impact on birds; they quickly returned to devastated areas except where trees had been killed (Manuwal *et al.* 1987). Leaf-feeding caterpillars of Lepidoptera ingest ash lying on the leaf surface, and those feeding on hairy leaves that retain ash suffered significantly greater mortality than those feeding on grasses from which tephra tends to be shed. Pyke (1984) discussed the effects of ash on the local mammals. Many of the small mammal species that survived the event are fossorial (Andersen and MacMahon 1985) and in May 1980 they had the additional protection of snow cover.

Among terrestrial vertebrates, an early pioneer of the revival on Mount St Helens was the pocket gopher, *Thomomys bulbivorus*. This small burrowing rodent feeds on plant roots, and Pendick (1996) noted that this lifestyle has features of considerable significance in the context of the Mount St Helens eruption. First, in its underground burrow, the animal was able to survive the blast and other immediate eruption effects. Second, its food would have been less affected by the eruptive activity than that of animals feeding on above-ground parts of plants. Third, in its burrowing activities, the gopher brings buried organic matter to the surface, thus working over the soil and preparing it for any returning plant seeds.

Time of day was also a factor in the Mount St Helens event. Had it occurred some 6 hr earlier than 08:30 hrs, many small mammals might have been out of their burrows, exposed on the surface. MacMahon *et al.* (1989) found that 7 years after the eruption, six of the presumed 32 resident small mammal species had returned to the zone of pyroclastic and debris flow, 15 had reoccupied the areas of tree blowdown and 22 were found in the area of tephra fall.

In intensive, if intermittent, studies on Anak Krakatau in the 1990s it was possible to identify those species of resident land birds, mammals, reptiles and butterflies that persisted after a 3-year episode of almost continuous Strombolian/ Vulcanian activity (Thornton *et al.* 1989, 2000). Species that persisted were predominantly those known to be early colonists following three catastrophic volcanic situations: Krakatau's 1883 eruption, Anak Krakatau's emergence in 1930, and Anak Krakatau's self-sterilization in 1952. It may be that early colonists have attributes enabling them to tolerate volcanic disturbance better than later colonists, or they may simply recolonize more rapidly after local extinction.

Lotic systems

Lotic systems (flowing fresh waters) that are not affected throughout by an environmental disturbance can recover quite rapidly when the disturbance ends. In the mid-1900s the Thames had become a 'biological desert' by the time it passed through London, but quite soon after proper supervision of London and the Thames catchment area was put in place, it was possible to catch fish from the London bridges. Systems in running water recover from damage quickly as their substrate is continually renewed; rivers and streams cleanse and renew themselves, if given the chance.

So too on Mount St Helens. By 1993, some 13 years after the eruption, trout were again seen in Spirit Lake, presumably having recolonized from feeder streams that were relatively unaffected. Stream systems recovered much more rapidly than those of the surrounding land habitat. Gersich and Brusven (1982) found that the aquatic larvae of stoneflies, caddis flies and mayflies withstand the effects of accumulated ash on their exoskeleton well, at least in the short term, as the ash deposits quickly dissipate in water. Within 5 years of the eruption invertebrate faunas of even drastically disturbed streams had made a good recovery, and in 10 years had reached 80% of their original diversity (Anderson 1992). Five years after the eruption several fish and amphibian species that had been locally extirpated had returned. Edwards (1996) noted that amphibians proved to be surprisingly resilient, returning even to the pyroclastic zone, with some populations recovering quickly.

CHAPTER FOUR

Life returns: primary colonization of devastated surfaces

One of the central themes of ecology is the origin of communities. This theme comprises a number of subsidiary questions, such as how primary plant and animal colonists become established on abiotic substrates, and how the various links and dependencies that go to make up a functioning community arise, and in what order; in fact, how community complexity develops and whether there are any 'rules' underpinning this. The natural reassembly of a biota following an extirpating eruption provides biologists with a rare opportunity to monitor the process, if they are lucky, from the beginning, allowing them to attempt to answer the first of these subsidiary questions.

There have been a number of studies of the recovery of areas that are not in themselves islands, which have been devastated by fire, flood, clear-felling, or by being covered with ash or lava. In such cases, the reservoir of species for recolonization is contiguous with the devastated area, so that it is sometimes difficult to distinguish true immigrants from the many transients. For the same reason, the ease with which individuals can enter the devastated area from adjacent unaffected species pools, recolonization occurs so readily that species extirpations are often precluded, or may occur without being noted by observers. Commonly, it is not known whether any remnant biota persist in the devastated areas – for example, as seed banks in the soil.

Primary colonists commonly have traits that maximize dispersal but allow growth on dry infertile substrates. Miles and Walton (1993) noted that the nature and direction of early plant succession (Fig. 4.1) on Mount St Helens reflects only the order of species arrival by predominantly aerial dispersal. Long-lived perennials or woody shrubs or trees with light, wind-dispersed seeds and slow growth are common among primary colonists (Table 4.1). Growth form and growth rate must also be appropriate to the substrate's supply of water and minerals. In general, trees are common primary colonists in rocky situations, biennial or perennial herbs on gravels, and grasses on silty substrates. Secondary pioneers generally have high growth rates, and in many volcanic situations, such as Krakatau, Taal and La Soufrière, they follow the primary colonists. The development of stands of woody plants can develop only when there are adequate nutrients (Shiro and del Moral 1993).

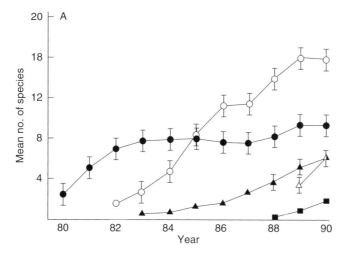

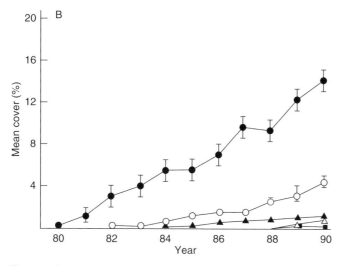

Figure 4.1 Mount St Helens: (A) mean species richness of plants and (B) mean percentage cover of vegetation (with standard errors) in 250 m^2 permanent plots on different regions after the 1980 eruption (years along abscissa, shown by last two numbers only). Open circles, lahars adjacent to intact vegetation; filled circles, scoured areas on high elevation ridge; filled triangles, blasted area on old substrate; open triangles, Pumice Plain; filled squares, Plains of Abraham (both are isolated pumice deserts, relatively bare before the eruption). (After del Moral and Bliss 1993).

Primary succession occurs on substrates that are purely mineral, without any binding of the mineral particles that would stabilize the soil surface, inhibit erosion and facilitate colonization. There can be no seed bank in such new ground, of course, no 'memory' of a previous flora, for there was none. The

Table 4.1 *Plant recolonization at Mount St Helens: five leading dominants in each study area; numbers are species rank order in respective samples determined by percent cover*

Species	Tephra	Lahar	Ridge	Pumice Plains[a]	Plains of Abraham[a]
Agrostis diegoensis	1				5
Anaphalis margaritaceum			2	1	1
Epilobium angustifolium			3	2	2
Eriogonum pyrolifolium		5			
Hieracium albiflorum			5	3	4
Hypochoeris radicata		4	4	4	3
Lupinus lepidus	2	2	1		
Penstemon cardwellii		3			
Phlox diffusa	3				
Phyllodoce empetriformis	4	1			
Polygonum newberryi	5				
Saxifraga ferruginea				5	

[a] Pumice Plains and Plains of Abraham are both isolated pumice deserts, relatively bare before the eruption.
Source: Del Moral and Bliss (1993).

assembly of species is dependent only on the chances of arrival and establishment of newcomers. Thus it is unlikely that successions in different places at the same time or at different times in the same place will be the same. Nevertheless, in contrast to Mount St Helens, some determinism is clearly evident on Krakatau, and is provided largely as a result of the island situation and the early establishment of a substantial proportion of seaborne propagules of very widely distributed species, neither of which is applicable in the Mount St Helens situation.

After the 1944 eruption of Vesuvius, colonization of scoria (loose rubble and ash mostly larger than 1 mm diameter) was much faster than colonization of lava (Mazzoleni and Ricciardi 1993). The life forms of laval colonists were chamaephytes (shoots bearing dormant buds on or near the ground) and hemicryptophytes (plants with dormant buds in the soil surface, aerial roots surviving for one season only). On tephra deposits the flora comprised, in addition, therophytes (annuals, seeds dormant in unfavourable season), phanerophytes (with aerial dormant buds) and geophytes (with underground dormant tubers, bulbs or rhizomes). On scoria, the main factors affecting the composition of the vegetation were found to be soil particle size and soil stability, in that order. *Rumex scutatus* was an early pioneer on coarse mobile scoria and pumices, and became dominant, although rare in the surrounding region. On steeper slopes it facilitated the establishment of other species like *Bromus erectus*, found growing within the clumps of *Rumex*.

Hendrix and Smith (1986) have studied the revegetation of a barren area formed following the 1968 eruption and summit collapse of the dry, 1494-m-high Galapagos island volcano Fernandina (Narborough), which is undisturbed by human activity. An area of 25 km^2 of vegetation was obliterated in 1968 by being covered with a layer of tephra 8 m thick; the area is now known as the 'barren zone'. Whereas lava flows mostly remain barren for more than 100 years, tephra on Fernandina's rim was revegetated after less than 10 years. In a 1984 eruption a large part of the rim collapsed into the caldera but only about 1 cm of tephra was deposited on the barren zone. Land iguanas (*Conolophus subcristata*) assisted recolonization in this zone by providing more suitable microsites by their burrowing activities and by depositing seeds, particularly of the important shrub *Solanum erianthum*, ingested outside the barren area, in their faeces. Nevertheless, by 1984 plant cover ranged from 0.1% to 1.1%, the three commonest species being *Sonchus oleraceus* (a cosmopolitan annual composite weed), *Aristida repens* (an endemic annual weedy grass) and *Eragrostis prolifera* (a widespread grass that can reproduce vegetatively once established). Water run-off channels and collapsed iguana burrows were major sites of colonization. As on Mount St Helens, erosion of tephra was important in re-establishment of plant communities, but this was much slower on Fernandina, and survivors were important as pioneers.

Nitrogen

Primary successions on volcanoes are on substrates that are usually unweathered, with low organic and nitrogen content and often with poor water and phosphorus availability. For succession to proceed, a pool of nitrogen is essential. The necessary nitrogen may be derived from the atmosphere if plant cover reduces wasteful rain run-off; from cyanobacteria; in some cases, as at Paricutin, from ammonium chloride in fumaroles; and from wind- or water-borne material, which averaged 4 mg/m^2 per day on Mount St Helens, amounting to over twice that in places.

Sprent (1993) noted that at maximum level nitrogen fixation could support the growth of some angiosperms, but it is not known how much nitrogen is fixed and how much is made available to subsequent stages of succession. Both legumes and non-legume nodulated plants are rare in early stages of primary succession, probably not because of the absence of their endophytic symbiotic microbes but because their seeds, necessarily rich in starter nitrogen, are too heavy for rapid or long-distance dispersal. However, cyanobacteria, including

nitrogen-fixing forms, were the first colonists of Vesuvius after the 1944 eruption and of Rakata, Surtsey, La Soufrière and Paricutin.

On Surtsey, the soil microbial community plays a vital role in the accumulation
of nutrients within the soil, which is essential for the primary succession from
bare soil to plant cover. Bare sandy tephra soils on the island have extremely low
organic matter content but plant growth is stimulated when nutrients become
available, thus retaining the introduced nutrients within the system for subsequent use in plant growth. During the succession to plant cover, microbial biomass and activity increased significantly following the establishment of seabird
colonies, and fungi came to dominate the microbial communities.

In the highlands of La Soufrière (1456 m), where persistent cloud maintains
hydration, cyanobacteria provided a gelatinous substrate for the colonization of
nitrogen-fixing lichens. Similarly, on Rakata they facilitated the colonization of
ferns, on Surtsey of mosses, and on Paricutin first of lichens and mosses and,
after 3 years, of ferns.

Lichens appeared on the Vesuvius lava after 10 years, seed plants not until
14 years after the eruption, and even in 1989 (45 years after the eruption), the
vegetation of most surfaces was still at the lichen stage.

In later succession on Rakata, after the fern cover had modified the soil and
conditions sufficiently, wind-dispersed seed plants followed. A third of the sea-
dispersed seed plants that colonized were legumes, 30 of their 34 species being
known to possess root nodules. The tree *Casuarina equisetifolia*, an important
pioneer of volcanic ash substrates on Krakatau, is also the most important
non-leguminous plant to fix nitrogen (see Chapter 8).

One of the early plant pioneers on Mount St Helens following the great 1980
eruption, appropriately enough, was fireweed (*Epilobium angustifolium*), well
known for its ability to colonize after fire by means of wind-borne seeds.
Another was the lupin (*Lupinus lepidus*), a legume, which has the added advantage, for a primary pioneer, of root nodules containing bacteria that fix nitrogen
from the air, enabling the plant to flourish without the need for nitrogenous
compounds in the soil (Halvorson *et al.* 1991). Where the roots of buried trees
were exposed by the rapid rain erosion following the eruption, they resprouted,
stabilizing the ash slopes.

Kjeldahl-extractable nitrogen within the organic fallout of arthropod bodies
on post-eruption Mount St Helens (see below) decreased by 25% in the first
17 days but was then fairly constant for the following 2 years. Edwards (1987)
presumed that this nitrogen was locked in the chitin and sclerotized protein of
cuticular structures, which remained more or less intact. In contrast, phosphorus in the fallout declined by 73% in the first 17 days, and by a total of 83% over
the whole 2-year period. Edwards' group estimated that for the years 1983–85
the mean net seasonal (150 days) input of nitrogen was 81.8 mg/m^2, and for
phosphorus 5.5 mg/m^2.

Table 4.2 *Colonization of Surtsey by mosses, 1967–72*

(a) Diversity	1967	1968	1969	1970	1971	1972
Number of species	2	6	7	18	40	66
Cumulative number of species		6	9	18	41	72

(b) Abundance[a]	1 quadrat	2–5 quadrats	6–25 quadrats	25 or more quadrats
Number of species, 1972 (64)	24	17	11	12

[a] Presence in number of 100 × 100 m quadrats, collectively covering whole area of island.
Source: Fridriksson (1975).

Colonization of lava flows

New lava flows are of course completely devoid of soil, fully exposed to the elements, and subjected to temperature extremes. However, where there is adequate moisture (as on Surtsey) colonization by microorganisms, mosses and lichens seems to occur almost immediately. Temperature, humidity, the nature of the substrate, and the size and proximity of the reservoir of potential colonists all determine the primary succession, which thus varies according to different primary situations.

New lava flows on Iceland, as on the Canary Islands, are colonized first by lichens, and yet it was not until 7 years after the 1963 eruption that the first lichen was found on Surtsey lava (Fridriksson 1975). On Surtsey, where humidity is quite high, nitrogen fixation by cyanobacteria was shown to be important (Henriksson and Rodgers 1978). Mosses were the dominant pioneer plant colonists, the first two species being seen on the lava in 1967; by 1970 there were 18 and by 1972 an astonishing 66 species (Table 4.2).

Richards (1952) referred to reports by Vaupel (1910) and Rechinger (1910) on the revegetation of lava flows on the island of Savaii, Samoa. Vaupel found that lava flows that were already 50–100 years old in the 1850s and were then said to have been covered with ferns and small bushes, by 1906 supported a well-developed closed forest above a dense undergrowth of mosses, orchids and grasses, the trees being rooted in cracks or on soil accumulated on the flow's surface. The forest comprised giant trees of *Rhus simarubifolia*, *Calophyllum speciosum* and species of *Eugenia* and *Ficus*, and epiphytic ferns and species of *Freycinetia* grew on tree trunks. Rechinger described the vegetation of young lava flows as consisting chiefly of low bushes with hard leathery leaves, such as *Morinda citrifolia*, and species of *Gardenia*, *Fragraea* and *Loranthus*, with a few grasses, the parasite *Cassytha filiformis*, and the pteridophytes *Nephrolepis rufescens* and *Lycopodium cernuum*. Very young flows, which originated in 1905, were still devoid of vegetation in 1906–07. Thus a tree cover was found some one and a half to two centuries after the eruption that produced the flows.

During the 1943–52 eruption of Paricutin in Mexico, primary succession had begun on lava flows by 1950, 7 years from the beginning of the eruptive episode. Ten years later there were 33 species of plants, and the shrub and herb growth was dense in places (Eggler 1963). Colonization of the cinder cone was similar to that on lava; the plant pioneers on lava flows, hornitos (small laval stacks from upwelling lava) and cinder cones were species known to have wind-dispersed disseminules, except two with fleshy fruits that were probably dispersed by birds.

On Lanzarote in the Canary Islands, lava flows were colonized primarily by lichens and mosses if they were not far from *kipukas* and not within 2 km of the sea (Kunkel 1981). In contrast to Surtsey, for example, on the dry Lanzarote lava close to the sea Ashmole *et al.* (1990, 1992) believed that input of marine bacteria plays a significant role in the maintenance of the lava ecosystem, and significant populations of heterotrophic bacteria were recorded. On most barren areas of the Canaries, however, the most important resource for animals is organic aerial fallout (see below).

Colonization of ash fields and tephra deposits

In general, tephra and ash deposits are colonized much more quickly than lava flows.

Although a year after the devastating 1902 eruption of La Soufrière, the young ash could not support plant growth, by 1907 humus was being formed in some places (Sands 1912). In 1912 a number of plants were established, most having arrived as wind-borne seeds or spores. Among them were three species that were important pioneers on Anak Krakatau (see below), the fern *Pityrogramma calomelanos*, the clubmoss *Lycopodium cernuum*, and *Eupatorium odoratum*, as well as grasses and sedges. After about 30 years of weathering, leaching and accumulation of organic material, the volcano's soils were comparable in organic matter and nitrogen content with cultivated soils on the island, and were probably capable of supporting climax vegetation. The vegetational successions seen at various altitudes (Beard 1976) correspond well with those recorded at various altitudes on Rakata, Krakatau's remnant. In La Soufrière's tropical rainforest zone the earliest tree colonists were heliophilous (sun-loving) wind-dispersed species which are often the dominants of secondary growth on abandoned cultivated land. At higher altitudes the seral communities resembled vegetation types that are usually found very much higher on mountains: elfin woodland, paramo and alpine tropical grassland.

After the 1944 eruption of Vesuvius, the first colonizers in the primary succession on coarse scoria were woody species, annuals being unimportant. The main factors affecting the vegetation of scoria were found to be soil particle size, followed by soil stability. *Rumex scutatus* was an early pioneer on coarse mobile scoria and pumices, and became dominant, although it was rare in the surrounding region.

A 'microsite' is the place at which an individual plant grows. Microsite patterns affect plant colonization in the harsh primary successional environments present in the immediate aftermath of catastrophic eruptions such as those of Kula (Turkey), Tarawera (New Zealand), Fernandina (Galapagos), Usu (Japan) (Shiro 1991), and Mount St Helens (del Moral and Bliss 1993). Their effect, however, is complex. In an interesting series of experiments on Mount St Helens, del Moral and Bliss (1993) compared the establishment and growth of plants from seeds that were planted at different naturally occurring microsites with those from seeds naturally dispersed into different artificially constructed microsites, over two seasons. Somewhat surprisingly, the microsites at which most seeds arrived naturally were not those at which there was peak establishment and seedling growth; moreover, microsites with the highest establishment rate of seedlings when artificially seeded did not have the seedlings with greatest growth. Further, microsite colonization patterns changed from one year to the next, when the weather and climate differed, and thus presumably so did microsite amelioration and seed rain. Thus on Mount St Helens the presence of a colonizing plant species in the early stage of primary succession depends on interaction between the seed-trapping ability of the microsite, the density and composition of the seed rain, and the ability of the species to become established under the conditions obtaining there (del Moral and Bliss 1993, 1987).

Extrinsic energy

Energy usually enters an island system as sunlight, passing through the photosynthetic pathways of green plants to animal herbivores and thence to carnivores, parasites and scavengers. The general textbook sequence of colonization is: plants, herbivores, carnivores. In this context many spiders have been regarded as exceptional carnivores; they arrive early because of their good dispersal as ballooning juveniles, and can survive on food carried to the island, like themselves, in the air. But they may not be so exceptional.

Airborne organic fallout

There is now a considerable body of evidence to show that in many instances the first organisms to colonize pristine or devastated areas are not primary producers, to be followed by herbivores and then carnivores and scavengers. Allochthonous (deriving from outside the area) wind-borne arthropod fallout can be a major source of energy in habitats where primary productivity is low or absent, such as alpine snowfields, oligotrophic lakes, and deserts (Edwards 1987, 1988), and newly exposed glacial moraines in the high Arctic (Hodkinson *et al.* 2001), as well as areas of new ash fall or new lava flows.

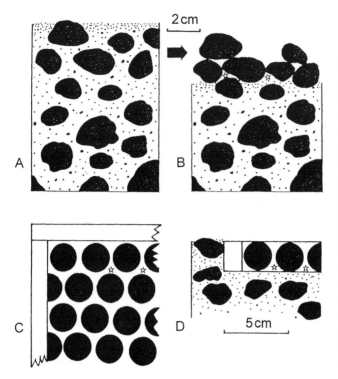

Figure 4.2 Mount St Helens: pattern of desert pavement after eruption, and a technique for monitoring invertebrate colonization from aerial fallout (see text). A, profile of early stage of desert pavement; B, subsequent development by wind removal of fine surface particles to leave surface stones with gaps in which arthropods (indicated by stars) can retreat after arrival; C, golf balls, shown here slightly separated, as an analogue of B, were packed into a wooden frame with a mesh floor and placed in the pavement (profile as in D) as a means of sampling arthropods. (After Edwards 1986.)

In the volcanic Canary Islands, the arthropod fallout forming the basis of a food web was best analysed on the snowfields of Tenerife's 3718-m-high volcano Pico Tiede, where the Ashmoles found 37 arthropod families, 17 of them Diptera, at densities reaching 18 individuals per square metre, with aphids and psyllids making up 53% of the total numbers (Ashmole and Ashmole 1988, Ashmole *et al.* 1990).

Organic aerial fallout was an important source of nutrients in the recoloniza-tion of Mount St Helens, following its great 1980 eruption. Edwards and his colleagues studied the terrestrial arthropods that were carried in the air to the blast zone, an area devastated by the eruption. Their sampling sites were at least 3 km from potential source areas of any kind, and at least 10 km from relatively unaffected source areas, so that their collections represented disper-sal over considerable distances. On their first visit, 2 months after the eruption, it was clear that wind-borne plants and arthropods were arriving from outside the area in appreciable quantities; fragments of plants and animals were found aggregated in natural micro-refugia – sheltered crannies and surface cavities. Edwards' group subsequently used wet pitfall traps, as well as specially designed fallout collectors (made from old golf balls (Fig. 4.2); we later modified this technique on Long Island by substituting the far lighter and cheaper table-tennis balls (Edwards and Thornton 2001)) and dry pitfall traps as analogues of the natural micro-refugia. Representatives of 140 families of arthropods from 17 orders were collected, the most diverse orders in the insect fallout being Diptera (42 families), Coleoptera (30), Homoptera (18), Lepidoptera (15) and Hymenoptera (9), and there were nine families of spiders. Diptera, Hymenoptera and Homoptera dominated as far as numbers were concerned. For the 2 years immediately following the eruption, chironomids and culicids were the major dipteran components, and this was probably a reflection of the increased productivity of freshwater bodies as a result of being organically enriched by the eruption (Edwards 1987). The organic fallout on to the surface of this blast zone averaged 5 mg/m^2 per day for both arthropod and non-arthropod material and reached summer maxima of 18 mg/m^2 and 26 mg/m^2 per day, respectively. Edwards (1996) estimated conservatively that from 2 to 10 mg dry weight of fallout per square metre per day represented non-resident arthropod bodies, and in the summer the total organic (arthropod and plant) fallout was 5–15 mg dry weight per square metre per day.

In polar semi-desert and tundra heath on the island of Spitzbergen in the high Arctic (latitude 78° N), Hodkinson and colleagues found chironomid midges, the dominant proportion of the total flying insect biomass, in densities beyond those that normally would result from local larval populations (Hodkinson *et al.* 1996), and which represent a substantial input of nutrients. It was estimated that the peak input rate of nitrogen, for example, was two orders of magnitude greater than estimates of rates of nitrogen fixation in Arctic habitats (Hodkinson *et al.* 2001).

Lindroth *et al.* (1973) suggested that an island may tend to receive more aerial fallout than its area would imply, because of a concentration of airborne material in its lee and updraughts on hot days sucking air towards the island from all directions. The latter point would apply particularly to tropical islands, and especially to those that are volcanically active. On Anak Krakatau in 1985 fallout traps, set 1.5 m above ash-covered lava flows, in 10 days collected more than 70 species of arthropods, including spiders, flies (the most numerous component), mirid and lygaeid bugs, aphids, cicadellids, psocopterans, moths, and beetles of ten families (Thornton *et al.* 1988a). Twenty individuals were collected per square metre of trap per day and it was estimated that at least half a million insects arrived on the 2.34 km^2 island in the 10 days of trapping.

Exploitation of fallout by pioneer colonists

In several studies of new volcanic substrates, and in a study of a newly exposed glacial moraine, the primary colonists are carnivore–scavengers that are members of guilds that exploit organic material arriving in the area as wind-borne fallout on to the barren substrate. These guilds of secondary consumers may themselves form the base of an island food web. The tapping of energy flowing through ecosystems outside the island, through the conduits of these guilds as well as through land-based consumers of marine resources such as fish, sea snakes, crabs and turtle eggs, short-circuits the usual plant–herbivore–carnivore energy pathway and enables the establishment of some animals without the need for the prior establishment of plants.

On bare lava fields of the Kilauea volcano on Hawaii, the fallout-exploiting community comprises a lycosid spider, an earwig, a mantid and a crepuscular flightless nemobiine cricket, *Caconemobius fori*, which dominates this guild of secondary consumers (Howarth 1979, Howarth and Montgomery 1980). Most species of the Hawaiian endemic genus *Caconemobius* are rock-dwellers on cliffs, lava flows and lava caves, and Frank Howarth believed *C. fori* to be closely related to a cave-dwelling species. He suggested that the entire aeolian community is a special subset of a more extensive guild, on the five main Hawaiian islands, of arthropods living below the surface in the network of caves and interstices of lava, and dependent on input of organic material from the surface. Besides the crickets, this subterranean community includes lycosid spiders, collembolans, amphipod and isopod crustaceans, a pseudoscorpion, centipede, millipede, earwig, cixiid bug and carabid beetle.

Near the summit of 4205-m-high Mauna Kea on the island of Hawaii, Howarth (1987) found an unusual, long-legged, flightless lygaeid bug, *Nysius wekiuicola*. The many Hawaiian species of *Nysius* are seed-feeders, but the Mauna Kea species is a scavenger–predator, having presumably evolved this habit as an adaptation to the high barren habitat. On the heights of Haleakala (3055 m) on the island of Maui, Howarth collected a strange flightless moth, *Thyrocopa apatela*. Its larvae make silken webs under stones and feed on wind-borne debris, mostly fragments of dead leaves of *Dubautia menziesi*, an endemic alpine shrub. It has evidently evolved the lifestyle of a kind of vegetarian spider! Lycosid spiders (wolf spiders) were also members of the guild on both islands.

After the 1980 eruption of Mount St Helens, the organic fallout was thought to contribute to the fertility of the pyroclastic flow and tephra substrates (Edwards and Sugg 1993), facilitating the early development of microorganisms and plants by the provision of nutrients from aggregations of arthropods, spores and seeds (Edwards 1996, Sugg and Edwards 1998). It was exploited by resident pioneer predator and scavenger guilds including carabid and tenebrionid beetles, spiders and neuropterans (Edwards and Sugg 1993, Sugg et al. 1994). Within the first two summers, 43 species of spiders had ballooned in from lowland areas to the west, and other colonizing pioneer predators and scavengers, carabids (ground beetles), arrived. Later, tenebrionids (darkling beetles) and neuropterans (lacewings) made their appearance. For all of these the fallout was shown to be an important source of nutrients (Edwards and Sugg 1993, Sugg et al. 1994). Carabids with aphids or nematoceran flies in their mouthparts were encountered frequently, particularly at night, and tenebrionid beetles scavenged on dead grasshoppers. Crows were also significant consumers of fallout during the day.

The first successful exploiters of this energy source on Mount St Helens, found in the spring following the eruption, were three carabid (ground beetle) species of the genus *Bembidion*. Species of this genus are generally restricted to disturbed, barren habitats, and are specialist, necessarily transient, pioneer colonists. Nineteen species of arthropods showed evidence of establishing breeding populations in the blast zone of Mount St Helens within the first three years. They included 11 species of carabid beetles, as well as other scavenging beetles. After 2 more years over 30 carabids and 125 spider species had arrived, making up about a quarter of the fallout-exploiting fauna and contributing about 100 individuals per square metre per year. About half the total numbers were lycosid spiders, and about a third were linyphiids, which made up almost half the species. By 1986, six species of these families had established populations on the surface of the pyroclastic flow deposits. In later years hemipterans (lygaeids and saldids), orthopterans (gryllacridids) and grylloblattids made their appearance. The gryllacridids were flightless and taken in pitfall traps in 1984; like phalangids (harvestmen) and a non-ballooning spider (an amaurobiid) taken in 1985, they probably walked into the area.

In barren areas of Anak Krakatau that were covered in ash and lava, but not in vegetated areas, we found a guild subsisting on the fallout of wind-borne invertebrates (New and Thornton 1988, Thornton *et al.* 1988a), that paralleled those discovered at Mount St Helens and on the island of Hawaii. Traps set on the barren ash surface caught representantives of a guild of scavenger–predators including lycosid spiders, earwigs, ants, a mantis and numbers of a nocturnal flightless cricket of the genus *Speonemobius*, of the same subfamily, Nemobiinae, as the cricket found by Howarth dominating this guild on Hawaiian lava flows. The parallel between this community and that on the island of Hawaii is remarkable (Fig. 4.3). On Anak Krakatau the guild of tertiary consumers of fallout on the ground included two migrant birds, the common sandpiper and grey wagtail. Aerial predators, such as two or three species of dragonfly, two swiftlet species, a swift and a swallow, took the fallout before it settled. These, in turn, were preyed upon by the Oriental hobby and perhaps also by the peregrine falcon when this replaced the hobby.

On Surtsey, which is about the same height and size as Anak Krakatau but more isolated and in a much harsher environment, colonization by pioneer animals exploiting airborne fallout has been much slower. Linyphiid spiders were seen alighting on the lava in the early years (Fridriksson 1975) and flies were the most diverse group of the island's early fauna, but by 1970 only three species of beetles were present.

Ball and Glucksman (1975) found that in inland high dry areas of Motmot, lycosids were already present in 1969, a year after the destructive island-forming

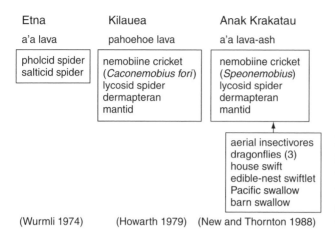

Etna	Kilauea	Anak Krakatau
a'a lava	pahoehoe lava	a'a lava-ash
pholcid spider salticid spider	nemobiine cricket (*Caconemobius fori*) lycosid spider dermapteran mantid	nemobiine cricket (*Speonemobius*) lycosid spider dermapteran mantid
		aerial insectivores dragonflies (3) house swift edible-nest swiftlet Pacific swallow barn swallow
(Wurmli 1974)	(Howarth 1979)	(New and Thornton 1988)

Figure 4.3 Patterns of the invertebrate community formed from aerial fallout on volcanic lava in several parts of the world. For each, the top box indicates the major 'core' taxa present in this aeolian community; for Anak Krakatau, immediate exploiters of this are also shown. A'a, rough broken lava; pahoehoe, smooth-surfaced or 'wavy' lava, (After Thornton 1991.)

eruptions ended, 'when, in spite of a careful search, no other life was found there'. They suggested that a major part of the spiders' diet may have been the fallout of aerial plankton sucked in following updraughts. On Motmot, the island that emerged from Long Island's caldera lake and whose colonization began in 1968 (Chapter 7), damsel flies and dragonflies, as well as Pacific swallows, *Hirundo tahitica*, may have exploited the falling aerial plankton before it reached the ground. All were present at the first survey, in 1969, and the swallow nested on Motmot in 1971. Earwigs and lycosids gradually spread until by 1972 they could be found over almost the whole island.

Two species of spiders are evidently the main exploiters of aerial fallout on Motmot. In 1999, a nocturnal ground-hunting wolf spider (Lycosidae), *Geolycosa tongatabuensis*, had very dense populations on the emergent island, presumably taking insects landing after they emerged fom the surrounding fresh water of Lake Wisdom. A tetragnathid spider was also very common, spinning webs near the ground between laval prominences and sedge tufts.

On new lava flows on Tenerife and Lanzarote in the Canaries (Fig. 4.4) the fallout-exploiters included an isopod, phalangid, centipede, collembolans, mites, seven families of spiders and five predacious and scavenging insect orders including an earwig, a predacious bug, a melyrid beetle and crickets (Ashmole and Ashmole 1988, Ashmole *et al.* 1990). Feeding on these, a guild of tertiary consumers included pipits and lizards.

On newly exposed moraines following glacial retreat on Spitzbergen, Hodkinson's group found that up to four species of linyphiid spiders were abundant and at high densities from the earliest stages of succession, where the unconsolidated fine particles of substrate were only partially stabilized by a thin crust of cyanobacteria, the community of soil invertebrates was rudi-mentary and vascular plant cover less than 1%. The number of spiders caught in water pitfall traps showed a strong positive correlation with numbers of

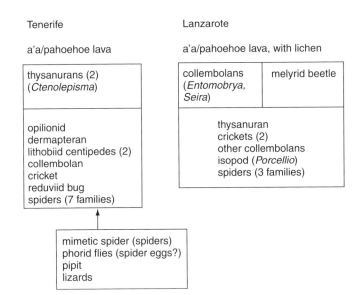

Tenerife

a'a/pahoehoe lava

| thysanurans (2) (*Ctenolepisma*) |
| opilionid dermapteran lithobiid centipedes (2) collembolan cricket reduviid bug spiders (7 families) |

| mimetic spider (spiders) phorid flies (spider eggs?) pipit lizards |

Lanzarote

a'a/pahoehoe lava, with lichen

| collembolans (*Entomobrya, Seira*) | melyrid beetle |
| thysanuran crickets (2) other collembolans isopod (*Porcellio*) spiders (3 families) | |

Figure 4.4 Patterns of the aeolian invertebrate community on lava on the Canary Islands; those on Tenerife and Lanzarote are broadly similar but differ in some details (Ashmole and Ashmole 1988).

chironomid potential prey (correlation coefficient 0.98, with less than one chance in 100 that this was fortuitous) (Hodkinson *et al.* 2001). Hodkinson's group noted that both density and number of species of spiders were often comparable at Arctic sites with those in much more productive, more temperate habitats. They suggested that the spider population played a number of crucial roles in the early stages of succession. Entrapment of wind-borne allochthonous prey items from outside the system by spiders ensures the retention of nutrients and their digestion of prey may accelerate nutrient release into a system where microbial decomposer activity is slight. Horizontal sheet webs covering the substrate entrap wind-blown debris and are themselves high in nitrogen and redistribute nutrients over the surface. They may also stabilize the surface and, through condensation, increase moisture availability there.

In all the above cases, organic fallout on to barren volcanic surfaces provides a resource for a guild of consumers (carnivores and scavengers), enabling them to become established on the ash and lava fields, free from the competition that would be a feature of more complex habitats. On Anak Krakatau the aeolian-based community proved to be the ecological analogue of communities on Mount Etna (Wurmli 1974), the Canary Islands (Ashmole and Ashmole 1988, Ashmole *et al.* 1990), Mount St Helens (Edwards and Sugg 1993), and, parti-cularly, on lava fields (but again not in adjacent vegetated areas) on the island of Hawaii (Howarth 1979). On Anak Krakatau, as on Lanzarote of the Canary Islands and the island of Hawaii, the scavenging/predaceous guild appeared to be adapted for life in areas of low primary productivity, and was absent or poorly represented in areas where productivity is provided by macroscopic terrestrial plants. On Lanzarote, for example, Ashmole and her colleagues found that the collembolans, flightless melyrid beetles, flightless crickets and isopod crustaceans that make up this guild were present in good numbers on recent lava close to the coast, but not on lava further inland or on older lava or *kipukas*. Fallout-exploiting communities appear to be characteristic of areas with a long history of sustained volcanic activity or with low primary productivity for other reasons (snowfields, glacial moraines), and may have evolved in situations where such environments have persisted over long periods of time (Thornton 1991), providing a good example of parallel evolution at the community level.

In spite of the obvious parallels between these guilds, of course there are considerable differences in their make-up. Thus, although a single carabid beetle was caught on Motmot in 1972 (Ball and Glucksman 1975), carabids, which were such important scavenger–predators in the early succession on Mount St Helens after the 1980 eruption, were not found on Motmot in 1999 despite extensive collecting. Nor were they important in the early fallout-exploiting community on Anak Krakatau. On the Canary Islands, although gryllids are present, Thysanura (bristle-tails) of the genus *Ctenolepisma* are important exploiters of fallout, one species evidently being something of a specialist, like the specialist fallout crickets of Anak Krakatau and Hawaii being confined to young lava flows (Ashmole and Ashmole 1988). In contrast, thysanurans were not important on Motmot, Anak Krakatau or Hawaii. The Canaries guild differs from others in another respect. Instead of two species of spiders dominating the community, as on Motmot, for example, on the Canaries' flows the spider component of the community is diverse; seven families are represented and there are specialist spider-eating spiders (Mimetidae) and spider-egg-eating phorid flies.

Finally, in certain circumstances some plants may have an ecological role broadly parallel to that of the animal guilds of airborne fallout exploiters discussed above. Michelangeli (2000) showed that in habitat islands on the sandstone summit of a Venezuelan mountain, carnivorous plants are unusually well represented. Such plants have a fairly rich source of insect food supplied by the summit updraughts, and, like the animal guilds discussed above, their activity brings nitrogen into a system that suffers from nitrogen deficiency.

PART III

The recolonization of devastated islands

Islands as areas for the study of community assembly

Compared with theoretical or laboratory studies on community assembly, studies of natural microcosms such as tree holes, or studies of the recovery of devastated continental areas, studies of the rebuilding of true island biotas on a macro-scale have both advantages and disadvantages.

Island studies are often more informative than studies of the recovery of continental areas simply because the devastated area is clearly defined and separated from its source, so that colonizing organisms can usually be clearly identified as such, and not confused with transients. Moreover, the encircling barrier reduces the likelihood of a 'rescue effect' (declining populations being 'rescued' from extinction by infusions of individuals from populations outside the area in question), thus allowing extinctions to occur and to be recognized. With both immigrants and extinctions clearly identifiable, turnover rates can be estimated, at least in theory, with some precision.

The most obvious disadvantage of such studies is that they are seldom possible. Destruction of an island's biota followed by natural reassembly is a rare occurrence, and if such natural experiments are to be exploited fully by scientists they must be recognized and acted upon quickly. The rarity of such events means that each must be analysed and studied; scientists cannot afford to let one pass unstudied. Each is fundamentally one of a small number of parallel 'replicate' natural experiments with considerable importance in helping to develop any general picture that exists.

Unlike experimental assemblies in which the substrate 'islands' (for example, glass slides or vials) can be made identical and only the variable under study (for example, colonization sequence) is altered in different trials, in the few natural situations available for study the islands are not identical. They have different sizes, heights, climates, geological ages and compositions, degrees of isolation (i.e. colonization rates) and degrees of disturbance, and carry biotas from different sources and of differing ages. They make very poor experimental situations. There is no control over the variables involved, and usually the analyses must be made by comparing and contrasting the results of natural assembly processes in a number of widely different situations. True replication is therefore impossible.

A further disadvantage of islands is the relative slowness of the process. Other things being equal, recovery of devastated isolates will be much slower than the recovery of equivalent areas of devastation that are contiguous with unaffected regions. In an island situation the new components of the recovering island land biota must first overcome the barrier of isolation – they must get there. In the normal island situation this is accomplished by crossing an aquatic (fresh or sea water) barrier within which most colonists cannot survive, let alone flourish.

The barrier does, however, have one advantage. It may be a conduit for extrinsic energy, and enables island communities to exploit an energy source unavailable to the pioneers of continental areas – organic flotsam.

Organic flotsam

A pair of ravens, *Corvus corax*, has been seen in summer on Surtsey regularly since 1965. Ravens are omnivores, and feed on carrion, exploiting resources such as washed-up carcasses that derive from outside the island itself. Sturla Fridriksson told me that up to 1999 the Surtsey raven pair had not bred on the island (they have not done so by 2005). Seabirds, the only other vertebrates and the first to land on Surtsey, also derive their sustenance from the surrounding ocean, feeding on carcasses washed ashore and sometimes bringing their catch on to land to feed.

Within 6 months of Anak Krakatau's emergence from the sea in 1930 a pioneer detritivore community consisting of large numbers of a collembolan, chloropid flies, and anthicid and tenebrionid beetles was exploiting seaborne organic debris. Both Bristowe (1931, 1934) and Toxopeus (1950) recorded the presence, at very early stages of the biotic succession, of members of a guild of scavengers and detritivores subsisting on organic flotsam, thus indirectly exploiting energy deriving from outside the island, in many cases from the sea itself. Shorebirds and kingfishers (which feed on crabs), fish eagles (which feed on sea snakes as well as fish), and the omnivorous large monitor, *Varanus salvator*, whose food includes organic flotsam, crabs and eggs of the green turtle, have together also acted as food-web links through which marine resources entered the island system.

Very few of the arthropod species that arrived early on Surtsey by air (see below) became established. Within 6 months of Surtsey's emergence from the North Atlantic in 1963 a strand community dependent on the algal, log and bird carcass flotsam had become established. Lindroth and his colleagues (1973) found that flies were an early group of colonists dependent on carrion but when there was no carrion they became extinct. By 1975 only three terrestrial invertebrates were thought to have colonized, a fly, a midge and a collembolan, all of them inhabiting washed-up carcasses.

On Motmot scavenging and omnivorous earwigs were amongst the earliest arrivals. They were probably not foraging on the beach for they gradually moved inland. An earwig, a species of *Labidura*, was eating other earwigs and lycosids. There was a strand fauna of small beetles, ants and bugs, again depending directly or indirectly on the input of organic material from outside the island. Ants among vegetation around Motmot's pond and staphylinid beetles and collembolans beneath a hardened algal crust at the pond margins had all arrived by 1972.

CHAPTER FIVE

Recovering island biotas: Volcano and Bárcena

In most recorded cases of the recovery of island biotas after natural extirpation, the earliest stages of recovery have not been monitored. In most instances there was no biotic survey of the island before the devastating event brought it to biologists' attention, often, although not always, because the event occurred so long ago. Thus the pre-devastation biota is usually unknown, or at least not known precisely. Sometimes, however, insight can be gained by assuming that the biota of the island before extirpation was similar to biotas now present on unaffected islands in the same region, if such 'control' islands exist. A comparison between these and the existing, reassembled biota of the affected island provides a fairly rough-and-ready estimate of the extent of recovery.

Another problem besetting such studies is that destruction of the previous community was often incomplete and the extent of destruction often only imprecisely known, so that the base from which recovery proceeded can only be guessed. One such case is Volcano Island in the Philippines.

Volcano Island, Lake Taal, Philippines, 1911

Taal Volcano, on the island of Luzon in the Philippine Archipelago, some 50 km south of Manila, is a low volcanic cone with a 21 by 13 km young prehistoric caldera complex – Lake Taal or Lake Bombon (about 300 km^2) (Fig. 5.1). Within the caldera lake lies an island, Volcano Island, which has an area of 25 km^2 and itself has a small crater lake.

Volcano Island is from 3 to 13 km from the lake-shore 'mainland'. A list of plants said to have been collected on the island in the late 1870s includes 236 species, but Brown *et al.* (1917) believed that the list must have been far from comprehensive. In 1910, however, the island's vegetation comprised a mixture of small fast-growing and fast-maturing trees typical of secondary growth, and grasses (mainly *glagah*, or wild sugarcane, *Saccharum spontaneum*). This is a common type of lowland vegetation known in the Philippines as *parang*, which follows removal of the original forest and fairly frequent fires.

The Taal volcano has been known to be dangerous since at least the arrival of the Spaniards in 1572; since then there have been 33 eruptions and the local

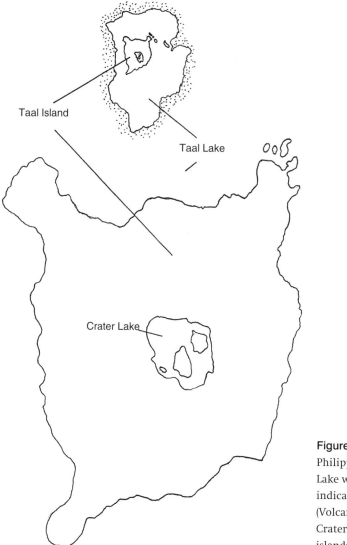

Figure 5.1 Taal Volcano, Philippines. Upper: Taal Lake with island indicated; lower: Taal (Volcano) Island, with Crater Lake and small islands present.

inhabitants have been repeatedly forced to evacuate. In 1910, however, there had been no activity from any of Taal's several volcanic cones since a fairly moderate eruption in 1874 covered the Volcano Island with 'ashes'. Many people were living on the lake-shores as well as on the island, where there were seven villages, and many cattle, water buffalo and horses grazed.

Taal erupted in January 1911. The Peléan eruption annihilated several villages and their human and animal inhabitants (with a loss of over a thousand human lives), and totally destroyed the vegetation over most of Volcano Island. Blast and pyroclastic flows (*nuées ardentes*) were responsible for most of the devastation, although the ash deposits were neither very deep (average about 30 cm, drifting to 2 m in places), nor very hot ('not much hotter than boiling water': Pratt 1911). A minor result of the eruption was the appearance within Volcano Island of a crater lake about 1 km in diameter, the surface of which was originally 70 m below the surface of Lake Bombon but rather quickly rose to about lake level. So there was an island (Luzon) in which there was a lake (Bombon), in which there was an island (Volcano Island), in which there was a lake.

Tropical rains can cause very rapid erosion of ash (see above, Chapter 2). After the 1964 eruption of Irazu volcano in Costa Rica, from half to a third of the newly fallen tephra had been removed from the upper slopes within a year of the eruption (Waldron 1967) and more than half the deposits round La Soufrière, St Vincent, a 1902 eruption identical in style to the one on Long Island but smaller, had been eroded to the ocean within 6 months. During the first rainy season after Taal's eruption, a few tree stumps were exposed on Volcano Island.

Gates (1914) surveyed the flora of Volcano Island three times between October 1913 and April 1914 (about 3 years after the eruption). The vegetation consisted of grasses, shrubs and young trees. Rather dense thickets of *S. spontaneum* were later believed to have arisen from surviving plants (Brown *et al.* 1917), although this species was later seen to be an obvious pioneer on Anak Krakatau, where survival was impossible. A few individuals of several trees seemed to have survived the eruption: *Trema orientalis*, *Morinda oleifera*, *Pithecellobium dulce*, *Semecarpus cuneiformis*, *Ficus indica*, *Eugenia jambolana*, *Cratoxylon blancoi*, *Sterculia foetida* and *Annona reticulata*.

The first plant colonists, near the shore, were water-dispersed, but inland a wind-dispersed grassland soon developed, which was invaded in turn, almost as soon as it was established, by bird-dispersed vines, shrubs and small trees. Gates (1914) assessed almost half the recovering flora (of 176 species) as having been bird-dispersed, 31 species (18%) as having arrived by wind, and 13 (7%) by water. Almost 3 years later, in January 1917 (6 years post-eruption), a further 117 plant species had colonized (Brown *et al.* 1917), making a total of 293 species.

Thirteen species that Gates had found 3 years earlier were absent. Volcano Island now carried *parang* vegetation, with *S. spontaneum* still dominant in the new community, and only 13 plant species were common and widely distributed. The shrub *Tabernaemontana pandacaqui* and the coarse herb *Blumea balsamifera* were very common. The most prominent tree in the lower ravines was a fig, probably *Ficus indica*, which was abundant on the mainland. The most frequent colonists were grasses, chiefly *S. spontaneum*, with *Themeda gigantea*, *Imperata cylindrica* and *Miscanthus sinensis*. Trees were scarce, the most common being *Acacia farnesiana*, which regenerates easily after fire and is a common early

colonizer of grasslands in the Philippines. The seeds of this tree are not adapted
for dispersal by wind or birds but the seed-pod floats. Individuals were scattered
all over the island, however, so that it is likely that some plants survived the
eruption. *Ficus indica* (= *hauili?*) was also fairly common, followed by the trees
Eugenia jambolana, *Trema orientalis*, *Tabernaemontana subglobosa*, *Morinda bracteata*,
Pithecellobium dulce and *Antidesma ghaesembilla*, all of which are common on the
mainland and distributed by birds.

The primary pioneers of the shore were *Canavalia rosea* and *Ipomoea pes-caprae*,
two creepers that were also pioneers on Krakatau and Bárcena volcano at the
other side of the Pacific (see below). Both Gates (1914) and Brown *et al.* (1917)
remarked on the presence of the latter in places at distances of 250 m from the
coast and at heights of 30 m. On Anak Krakatau in the 1980s and 1990s (see
below) both these species occurred at similar distances from the shore. There
was a considerable fern flora in the ravines of Volcano Island, the most prominent species being *Acrostichum aureum*, *Pityrogramma calomelanos*, *Nephrolepis
biserrata*, *Pteris vittata*, *Blechnum orientale*, *Onychium siliculosum*, *Shenomeris*
(= *Odontosoria*) *chinensis*, *Pteris quadriaurita*, *Cheilanthes tenuifolia* and *Adiantum
philippense*. The first six of these species were recorded on Rakata (Krakatau's
remnant) by the time of the first survey, in 1886, 3 years after Krakatau's
eruption, *S. chinesis* was present after 37 years, in 1920, and *P. quadriaurita* after
68 years, in 1951.

Brown and his colleagues assessed 157 of the 293 species (54%) of the 1917
flora as having been transported by birds across a water gap averaging about
6 km, in 6 years. It was thought that 83 species were likely to have arrived as
propagules within the alimentary tract of the birds, 14 possibly attached to
feathers by barbs or hook-like structures, and 60 as minute seeds which could
have been imported in mud on birds' feet or feathers. Sixty species of plants
(21%) were assessed as having been distributed by wind, 26 (9%) by water and the
remainder by humans or their livestock. Most of the plant colonists (87%) were
species with wide geographic distributions, 36% being pan-tropical.

Brown's group compared the early stages of revegetation of Volcano Island
with that of Krakatau (Brown *et al.* 1917). In contrast to the Krakatau case, no
cyanobacteria (blue–green algae) were reported on Volcano Island and although
there were 21 species of pteridophytes, ferns were much less prominent than
they evidently were on Krakatau after its 1883 eruption and did not grow on
newly formed substrate. Sequence of colonization by dispersal mode on Volcano
Island was broadly similar to that found on the Krakataus, with an early grassland (see below), but was much faster, perhaps because of the much shorter
distances from source populations. For example, closer source populations
would increase the chances of ingested seeds surviving the flight of a bird or
bat to the island, and in fact bird dispersal became important very much earlier
on Volcano Island than it did on Krakatau, where distances (from 13 to 44 km)

were much greater and near or beyond the limit of the retention time of many birds and bats. Since Volcano is not a marine island the typical strand formation which developed early on the Krakataus was lacking from Volcano, where the inland vegetation was the first to develop.

After 54 years of quiet, in 1965 the water temperature of the Volcano Island lake became abnormally high. Scientists at Taal's small volcano observatory warned of a forthcoming outburst, and there was a partial evacuation of the area. The volcano burst into renewed activity on 28 September, and in the next two days 190 people who had not left the area perished. The work of Moore (1967) and Waters and Fisher (1971) on this short, violent, hydrovolcanic erup-tion brought base surges into prominence in volcanological studies. These are ground-hugging, low-density, turbulent clouds that roll away at high velocity radially from the base of a volcano's overloaded, collapsing vertical eruption column. They seem to be produced particularly in eruptions involving interac-tion between magma and water. The 1965 Taal eruption began in Strombolian mode, a lava fountain beginning the construction of a cinder cone. However, following the entry of lake water into the vent through a crack, the eruption's character changed abruptly. Powerful horizontal blasts containing ash and frag-ments of volcanic rock surged with hurricane force from the base of Taal's column, moving over the surface of Lake Taal to its shores, engulfing fleeing boats on their way. Trees within a kilometre of the vent were snapped off or uprooted; those further away were defoliated and the ash-laden blasts scoured wood from their trunks to depths of up to 10 cm on the side facing the vent. The trees were not burnt or charred, simply sandblasted. The violent surges could not have been much hotter than 100 °C, for there were thick deposits of mud at the edge of the devastated area, about 6 km from the vent. The lower compo-nents of these base surges sandblasted objects up to 7 km from the vent. Deposits accumulated to a depth of 4 m within 2 km, and the volcanological station on Volcano Island was buried under 3 m of ash. This was the first time that ash dunes, more streamlined than sand dunes and including fragments too large to have been carried by normal wind action, were seen to have been produced from horizontally moving, particle-laden surges.

I know of no biological survey that was made since this 1965 event.

San Benedicto (Bárcena I), Mexican Pacific, 1952

The only other volcanic island to have been recently active and subsequently monitored by biologists, is Isla San Benedicto, of the Revillagigedo Archipelago, some 400 km from the Pacific (Baja California) coast of Mexico (Fig. 5.2). Consisting of several coalesced old volcanic cones, it is the youngest of the four Revillagigedo islands, and probably Late Pleistocene in age. The island has no fresh water.

Before 1952 the island's biota was oceanic in character. There was only one species of land bird, an endemic subspecies of the rock wren (*Salpinctes obsoletus*

Figure 5.2 Position of the Islas Revillagigedo.

exsul), and mammals, reptiles, amphibians and freshwater fish were completely lacking. There were 11 angiosperm species and the island was almost completely covered in grasses and low herbs, with the grasses *Eragrostis diversifolia* and *Cenchrus myosuroides*, and *Euphorbia anthonyi* dominating. The terrestrial invertebrate fauna was almost unknown – only a single species, a land crab, *Aegeocarcinus planatus*, had been recorded. Frigate birds (two *Fregata* species), with densities of from one to five birds per square metre, made up 95% of the numbers of seabirds, of which there were eight species.

At 07:45 hrs on 1 August 1952, in a valley on the southern part of the island, there was an explosive eruption which had the highest volcanic explosivity index of any oceanic volcano in the eastern Pacific. A new volcano, Vulcán Bárcena, appeared (Fig. 5.3). Within about 20 minutes the entire island (6.4 × 3.2 km) was covered in a dark cloud of ash and pumice, and within hours the layer of tephra was almost a metre deep and no plants were visible. Some 20 000 seabirds perished at this time, many of them covered with ash as they stayed on the ground, some on nests with eggs (their skeletons were exposed by erosion over a year later). By the next day the new volcanic cone was 335 m high. By mid September growth of the cone had ceased and the vent in the crater was capped by a small lava plug. In early November, however,

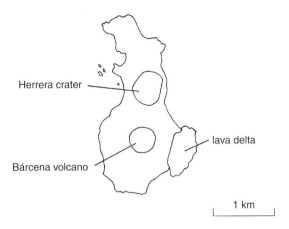

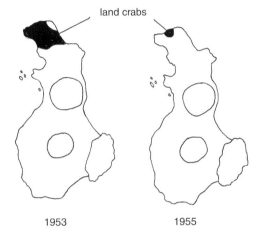

Figure 5.3 Bárcena Volcano
Upper: major features; lower:
distribution of land crabs
(shaded) in 1953 and 1955. (After
Brattstrom 1963.)

activity was resumed and two domes of irregular lava blocks were formed in the crater. On 8 December lava breached the flank of the cone and flowed from the base of Bárcena, and by the end of February 1953, when activity ceased, the flow had reached the coast and formed a lava delta which extended almost a kilometre out to sea.

Bayard Brattstrom (1963, 1990) has visited the island repeatedly to study the eruption's effect on the island's biota and the subsequent recovery. Only about half of the terrestrial biota, in terms of species, was found to have been destroyed. Five of the island's 11 plant species were lost: *Stenophyllus nesioticus*, *?Dodonaea viscosa*, *Ipomoea pes-caprae*, *Teucrium affine* and *Erigeron crenatus*. Six species survived by sprouting from old rootstocks or through germination of buried seeds after the ash cover had been reduced or lost by erosion. The survivors were the grasses *Eragrostis diversifolia* and *Cenchrus myosuroides* (only two individual plants), the sedge *Cyperus durides*, and the herbs *Aristolochia brevipes*, *Euphorbia anthonyi* and *Perityle socorroensis*. Three of these survivors did not recover well and became extinct within 9 years, but the other three, *E. diversifolia*, *C. durides* and *P. socorroensis*, have successfully recolonized the island (Brattstrom 1990).

The island has steep cliff sides and potential immigrants floating to the island would have been unlikely to make a successful landfall until two sheltered beaches were formed in the lee of the 1952 lava flow. Two colonizing plant species became established on these beaches. The first, in 1961, on the newly formed northern beach, was *Ipomoea pes-caprae*, the beach morning glory, one of the plants present before the eruption, and incidentally one of the first pioneers also on the Krakataus and on Volcano Island of Taal Volcano, on the other side of the Pacific. Since this beach was formed after the eruption, the species could not have resprouted from surviving stock. By 1971 it covered almost the whole of the beach, and in 1978 was invading the lava delta. In 1977 the plant was seen on another beach to the south, but did not persist there.

Brattstrom has told me that he thinks that *E. anthonyi* may have successfully recolonized the island after becoming extinct there. It was found in the nest of a red-footed booby (*Sula sula*) in the lava delta in 1971, when it was not growing anywhere else on the island. The plant could have been reintroduced from Socorro by these boobies or by frigate birds, which also use vegetation as nest material. Brattstrom also considers the establishment of *P. socorroensis* on the northern beach by 1971 to be a new invasion of the island.

The Revillagigedo islands Socorro and Clarión are about 60 km and 400 km, respectively, from San Benedicto (Fig. 5.2). Socorro is 1200 m high and the largest (at 15 × 15 km) island of the group, with over a hundred vascular plant species, and Brattstrom believed that it would be a major source for any recolonization of San Benedicto. However, he thought that establishment of immigrants would continue to be hampered by lack of rainfall and slow soil formation on the island, and thus poor re-establishment of plants as food resources for animals.

By 1971 the vegetation had recovered sufficiently to extend over almost all the northern part of San Benedicto.

Terrestrial invertebrates recorded after the eruption included spiders and many winged insects, including two species of grasshoppers, hippoboscid and muscid flies, honeybees and ants. The land crab species survived at the northern end of the island, where the ash layer was only a few feet thick, and has gradually extended its range since 1955.

The archipelago lies in the path of hurricanes, and 12 species of land birds have been recorded as stray visitors since Bárcena's eruption, six of them within 3 years of the eruption. The strays include the peregrine falcon, *Falco peregrinus*, and barn swallow, *Hirundo rustica*, two species that were colonists of Anak

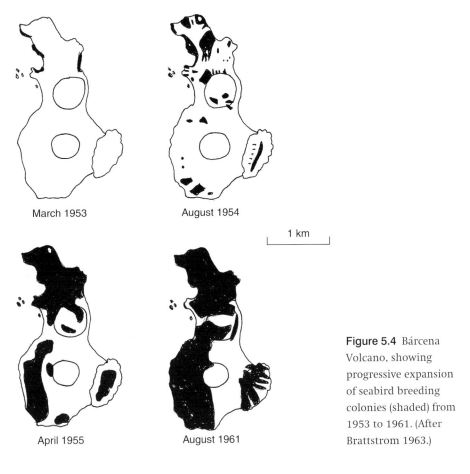

March 1953

August 1954

1 km

April 1955

August 1961

Figure 5.4 Bárcena Volcano, showing progressive expansion of seabird breeding colonies (shaded) from 1953 to 1961. (After Brattstrom 1963.)

Krakatau and Motmot in the western Pacific. None of the 12 has become established, although by 1982 the Socorro Island subspecies of the Socorro red-tailed hawk, *Buteo jamaicensis socorroensis*, and the Clarión Island subspecies of the Clarión raven, *Corvus corax clarionensis*, were occasional breeders on San Benedicto. A few rock wrens were seen to have survived in December 1952, but none has been seen since then and the subspecies is now considered to be extinct.

The total seabird population (Fig. 5.4) in November 1953 was estimated to have been about 1% of that in 1925 but although recovery of seabird populations was slow in the beginning, the eight seabird species that were on the island before the eruption recovered well. There are now populations of greater and magnificent frigate birds (*Fregata minor* and *F. magnificens*), wedge-tailed and Townsend's shearwater (*Puffinus pacificus* and *P. auricularis*), the red-billed tropicbird (*Phaethon aethereus*), masked and brown boobies (*Sula dactylatra* and *S. leucogaster*) and red-footed booby. The last species requires shrubs or trees for nesting and was the slowest seabird species to recover, although on San

Benedicto the birds breed on the rugged lava flows and cliff edges, as some of them do on Clarión Island, 364 km to the west. New arrivals, the blue-footed booby, *Sula nebouxii*, and brown pelican, *Pelicanus occidentalis*, have each been recorded from San Benedicto on just one visit, in 1953 (in November and March respectively), but the Laysan albatross (*Phoebastria immutabilis*) was reported nesting on the island in 1990 (Brattstrom 1990), the only new arrival since the eruption to become established.

In March 1953 seabirds were found only on the bare northern cliffs, but by November in the north of the island frigates occurred where the ash layer had been removed by erosion and boobies on the ash in areas sheltered from winds. By 1955 shearwaters had begun to dig their burrows in the ash at the top of Bárcena, and a few birds, mostly boobies, had begun to roost on the lava delta. By 1961 the seabird population had risen from several score immediately after the eruption to about 7000, and by 1971 to over 10 000, the large increase since 1953 being almost entirely accounted for by increase in masked booby numbers (perhaps significantly, a species that makes little or no nest and does not use vegetation). Ten years later this species, and the total seabird numbers, had declined considerably. Erosion of the soft pumice of Bárcena's crater and thermal dust storms had destroyed most of the shearwater burrows by 1971, but by 1978 the hard crust on the ash surface had largely disappeared and shearwater burrows again appeared on the crater and in other places.

The masked booby is now the most numerous bird species on the island, followed by the red-footed booby (since 1971) and Townsend's shearwater. Frigate birds, by far the most numerous before the eruption, when they comprised over 90% of the seabird population, have not built up large populations since. B. H. Brattstrom (personal communication) believes that the rapid increase in masked booby numbers since 1955 cannot be explained by natural population growth, and that the explanation must be immigration from elsewhere, with Clipperton, Cocos and the Galapagos as the nearest likely sources. The masked booby population declined markedly after 1971 (about 11 700 individuals) to 8600 in 1978 and 3400 in 1981, possibly as a result of lowered productivity in the waters around the islands associated with El Niño events.

There have also been changes from year to year in the nesting areas of the colonies of the main seabird species. These changes are unexplained but may be related to the rapid substrate changes associated with weathering and erosion.

There have been marked erosional changes on San Benedicto. The cone of Bárcena and other areas of ash are now furrowed with erosion channels 40 or 50 m deep, and Brattstrom (personal communication) believes that weather and the effects of erosion on the volcanic ash have had an important effect on the survival and recovery of plants, perhaps greater than the direct effects of the eruption. The beaches north and south of the lava delta have increased in size tenfold since 1961 and are now protected by berms of lava boulders. Ash has washed down on to the lava delta and in-filled almost a third of it, and seabird guano has added nutrients and stabilized the ash. Three new land forms, the lava delta and the two protected beaches, provide suitable habitat for seabirds and water-borne plants, and could be the sites of establishment of new animal and plant colonists (Brattstrom personal communication).

As yet no rats, mice or cats have been reported on the island in spite of occasional visits by fishermen, sightseers and scientists.

Three other case studies of island recovery – Thera, in the eastern Mediterranean; Long Island, north of New Guinea; and Krakatau, in Sunda Strait, Indonesia – are each accorded a chapter to themselves. All three suffered an enormous explosive eruption with a vast release of energy, and in each case the eruption had a long-lasting effect on human populations, those of Thera and Krakatau perhaps being of great cultural as well as historical significance. Thera and Krakatau are probably the two most famous cases of volcanic devastation of island biotas. The eruption of Thera, which occurred too long ago for its recovery to be monitored, nevertheless holds a special interest because of its probable important role in the demise of a remarkable Mediterranean culture. It has also been argued that a vast eruption in about 535 BC of 'Ancient Krakatau', the precurser of the islands whose recovery since the 1883 event has been monitored by scientists (Thornton 1996a), played perhaps an even more fundamental and far-reaching role in Earth's human history. Although there have been no claims of an equivalent historical importance for the eruption of Long Island in about AD 1645, the event is certainly an important part of the oral history of many linguistic groups in New Guinea (Blong 1982).

CHAPTER SIX

Thera, Santorini Group, Mediterranean

One of the oldest known cases of the recovery of an island's ecosystem from devastation is that of a volcanic island in the Aegean Sea, between Greece, Turkey and Crete in the eastern Mediterranean (Fig. 6.1).

Around 1620 BC, the island of Thera, in the small Santorini Archipelago of the Cyclades group, about 70 km north of Crete, erupted catastrophically. This eruption was not only very large and powerful by volcanic standards but also was probably one of the most influential eruptions in human history (perhaps after that of Ancient Krakatau in AD 475). It was an event of huge proportions that occurred in what was then a relatively small, developing, Western world, and thus took on a social, political and historical significance far transcending what even an eruption of this magnitude might have today. One view is that the decline of the Minoan civilization on Crete in the seventeenth century BC stemmed from the psycho-sociological effects of this eruption (known as the Minoan eruption), and there can be little doubt that, if not directly responsible, at the very least the eruption played an important role in the decline of that culture (de Boer and Sanders 2002).

In legend, the island is referred to as Stronghyli (round island) and was called Kalliste (most fair, best) by the Phoenicians, one of the first peoples to colonize it. Although officially named Thera, after the name given by the classical Greeks to its volcano, most people know it as Santorini, after its patron saint, Santa Irene. The last two names appear to be used interchangeably, without any system. Here, the name Santorini will be used for the present group of islands, and Thera for the largest island, which is the major remnant of Stronghyli's volcano.

The volcano is the only active one in the 20–40-km-wide Aegean volcanic arc, which runs along the boundary of the Aegean tectonic plate for some 500 km from the islands of Aegena, Methana and Poros, near the east coast of mainland Greece, south through the central Aegean, then sweeps east-north-east to the Bodrum peninsula on the south-west coast of Turkey (Fig. 6.2). The African tectonic plate is moving north under southern Europe at a rate of about 5 cm/year along this arc. The plate begins to descend along an arc south of Crete, and is at a depth of about 150–170 km below the Aegean arc, where the stresses

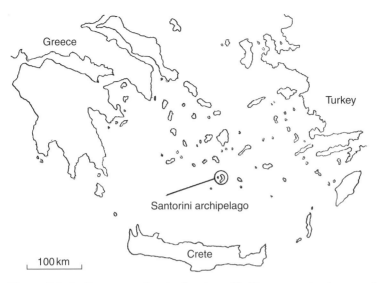

Figure 6.1 Outline map of part of eastern Mediterranean to show position of the Santorini archipelago.

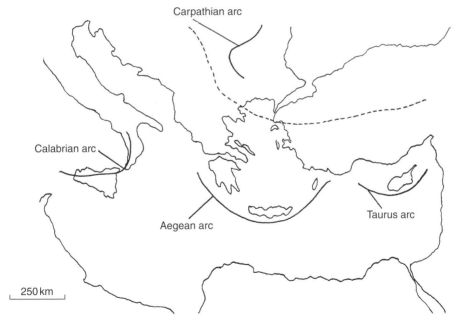

Figure 6.2 The major volcanic arcs in the eastern Mediterranean.

caused by its movement result in earthquakes centred at depths of about 170 km. As recently as 1999, destructive earthquakes in Greece and Turkey were dramatic testimony to such stresses. As well as earthquakes, volcanic activity is an accompaniment of these great earth forces, as the upper 100 km

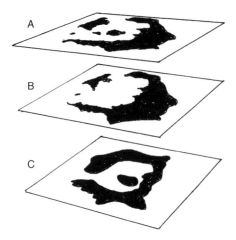

Figure 6.3 Santorini: the change in archipelago form with volcanic eruption: C, pre-eruption; B, post-eruption; A, present day. (After Freidrich 2000.)

of lithosphere is subjected to deformation and fracture, opening channels for the ascent of underlying magma. Volcanic activity on Kos, at the eastern end of the arc, began in the Miocene period, but along the rest of the arc activity commenced in the Pliocene, about 3 million years ago. The Santorini group, in the middle of this arc, is one of five volcanic centres related to zones of weakness characterized by frequent shallow earthquakes (less than 20 km deep) as well as volcanic activity. Volcanism on the Santorini group began in the late Pliocene but most of the activity took place in Quaternary, and in the last 200 000 years twelve volcanic centres have been active.

Freidrich (2000) provided a geological history of Santorini (Fig. 6.3). In the Pliocene, in the Santorini area, there was a non-volcanic island of eroded metamorphic rocks (now exposed on Profitis Ilias, the highest point of Thera, and several other places). A second island appeared in the Late Pliocene, when the first eruption broke the sea surface to the south west, in the area of Thera now called Akrotiri. About 500 000 years ago, when these two islands had fused under a mantle of volcanic rocks, another volcanic shield began to grow in the north of Thera, and by about 200 000 years ago volcanic products had produced a single large, oval island, Stronghyli, which included what is now Thera and most of what is now Therasia. The roof of the magma chamber of this volcano collapsed about 180 000 years ago to form the first caldera at the site, the Bu caldera, about 3 km in diameter. This became filled with eruption products from two other volcanic centres, until about 54 000 years ago there was a second collapse, forming a larger caldera, the Skaros caldera. Once again, lavas filled this caldera. Several volcanic vents formed an overlapping volcanic complex in the north, and in the south grew in a depression, probably open to the sea.

There was a huge ignimbrite eruption about 23 000 years ago in the north, which buried most of the island, including the northern part of the caldera, but

Figure 6.4 The Santorini archipelago, with major islands named.

the southern part remained as a 6-km-diameter caldera and was deepened. A shallow lagoon then appeared in the north (stromatolites were growing there some 15 000–12 000 years ago), which merged with the southern caldera, which was still open to the sea in the southwest.

It is then that Freidrich believes that a central 'Pre-Kameni' emergent island was formed, and grew within this caldera. The roof of the magma chamber beneath 'Pre-Kameni' eventually collapsed as it was emptied, and Freidrich believes that a concentric ringed structure may have been formed, a conjecture which has some relevance to a theory that this is the site of the legendary Atlantis.

On the basis of geological and archaeological evidence, Freidrich et al. (2000) and Freidrich (2000) have provided a reconstruction of Stronghyli prior to the so-called Minoan eruption of the second millenium BC (Fig. 6.3). The island was already ring-shaped, with a central submerged marine caldera open to the sea in the south-west, with an island, 'Pre-Kameni', within it. This configuration is similar to that of present-day Long Island, north of New Guinea, and its caldera islet, Motmot, except that Long's caldera is closed, and filled with fresh water, and the situation on Krakatau, except that Krakatau's caldera has three openings to the sea. Thus the Thera's present caldera, like that of Krakatau, is not the first at the site. The present-day configuration of the archipelago is shown in Fig. 6.4.

The pre-eruption environment

Freidrich (2000) noted that the break-up of the southern Aegean arc into a string of islands interrupted migration routes, led to the development of an endemic flora, and reduced floral diversity relative to that of the Greek and Turkish mainlands. A southern Aegean floral zone, which comprised Crete, Rhodes and Carpathos, served as a pathway between Europe and Asia, but the central Aegean islands, protected from competition from cold-resistant species that invaded the mainlands during the climatic changes associated with the Pleistocene glaciations, were able to to retain a large part of their Tertiary flora (Greuter 1979). However, during the Quaternary glacial epochs a vast amount of sea water was locked up in expanded polar ice caps, sea level was lowered by up to 150 m, and there were considerable periods when the above-water parts of the Aegean arc were much more extensive than they are today, so that plants, animals and people were able to spread into the region.

Before human occupation, Stronghyli was perhaps less arid than it is now. Although it may have had rather more trees than Santorini has today, it probably lacked browsing animals – the 18 000 BC eruption would almost certainly have eliminated any native mammals (Rackham 1990). Remains of the lentisc (*Pistacia lentiscus*) and therebinth (*Pistacia therebintus*) bushes, and date palms (*Phoenix theophrasti*), olive (*Olea europaea*) and tamarisk (*Tamarix* sp.) trees have been found, having been preserved by deposits of a pre-Minoan eruption dating about 50 000 BP (before present). *Pistacia lentiscus*, a member of the cashew family, is characteristic of the Mediterranean flora today, often in association with the therebinth, and is found on Santorini. The fossil olive leaves (which carried the olive white-fly, *Aleurobus oliviius*, as do olives in Crete today) are the oldest known in the Mediterranean, and evidence that this tree was not introduced by humans in historic times (Freidrich 2000). Remains of two species of palms killed in the pre-Minoan eruptions have also been found. One is the dwarf palm, *Chanaerops humilis*, a member of a genus now known only from the western Mediterranean, and the other is a date palm, *Phoenix theophrasti*, which is now almost confined to Crete and grows throughout Santorini. Most dwarf palm fossils were found in erosion channels on the volcano's flanks. Rackham (1990) conjectured that the vegetation of that time may have been an open woodland of lentisc and oak, with extensive steppes, an indication of a rather warmer climate than that of today.

Plant fossils found in 21 000 BP ignimbrite deposits filling old erosion channels on Therasia indicate that in the Late Quaternary Santorini was almost barren of vegetation. Wind and rain erosion reworking porous volcanic debris would hinder or prevent plant growth, and water is not retained in tuff (compacted volcanic ash) deposits, creating arid conditions so far as the plants are concerned.

Immediately before the Minoan eruption the environment would have been much as it is today – with a similar climate, extremely arid and probably treeless. At that time, as now, Thera was probably too low to cause clouds to shed their moisture as rain. The three volcanoes of the northern volcanic complex were no higher than about 400 m (Heiken *et al.* 1990), the highest point on the island probably being about the same as that of today, the 568-m-high limestone peak, Profitis Ilias. Thus Thera probably lacked, as it does now, the lush vegetation, including drought-sensitive plants such as holm oak (*Quercus ilex*), valonia oak (*Q. macrolepis*) and maple (*Acer orientale*), that often covers peaks over 800 m altitude on Mediterranean islands. As now, the lower vegetation may have lacked the leathery-leaved evergreen woody shrubs and small trees that comprise the Mediterranean *maquis*, instead consisting only of *phrygana* (grey–green aromatic short-lived undershrubs, such as *Cistus* and *Salvia*) and steppe (grasses and herbs with many annuals and bulbous plants).

On rural Santorini, excavations of Bronze Age levels have revealed remains of oak, pine, tamarisk, vine, olive and reeds (*Arundo donax*), and Rackham (1990) suggested that soils would have been more developed than those of today, enabling the cultivation of olives, vines and a wide range of crops as well as the growth of good pastures. During the long period of quiescence before the Minoan eruption, the inhabitants cultivated barley, figs, almonds, beans, lentils, chickpeas, a species of lupin, wheat, olives, and dates, and sage tea and marjoram were used (Diapoulis 1971, Rackham 1990). 'Faba' seeds (*Lathyrus clymenum*) were eaten (and are still a staple on Santorini) and reeds were used in the construction of shade awnings or screens. Lentils were evidently an important dietary item (Freidrich 2000). The madonna lily (*Lilium candidum*), strand narcissus (*Pankratium maritimum*) and crocus (*Crocus sativus*) also grew on Thera, the crocus gathered for its saffron.

The only evidence of wild (although non-native) animals before the eruption are hare bones, and Rackham, who believed that the 18 000 BP eruption would have wiped out any native mammal fauna, suggested that all the mammals of the Bronze Age and later must have arrived in boats. Bones of sheep and goat, with lesser numbers of pig and ox and a few of red deer, suggest typical Bronze Age Cyclades pasturage. The Akrotiri artists depicted cattle, deer, goats, bears, monkeys, Sömmering gazelles and a large predatory cat. The last three are depictions of exotic animals known from trade contacts with north Africa (an ostrich egg was also found fashioned into a vase).

The Bronze Age Therans evidently ate well. Trantalidou (1990), who has analysed animal remains on Santorini, found that animals were slaughtered at a younger age and that there was a higher percentage of cattle than at other Bronze Age sites. Both of these differences indicate a higher quality of life. It is likely that sheep were also kept for their wool and that cattle were used as beasts of burden as well as food (Cadogan 1990). In what was probably the main settlement and capital, Akrotiri, there seem to have been no fortifications, suggesting that the sense of peace and security that pervaded Minoan Crete at the time (see box) had extended to the Thera colony. Clearly, Thera prospered as a colony of Late Minoan I Crete, and Akrotiri was its prosperous port, but was not the only settlement. The colony's great success is unlikely to have had its basis in agriculture unless some special high-value crop was cultivated (Rackham 1990), and is more likely to have been the result of Akrotiri's good harbour within the sunken caldera and Santorini's geographical position as an entrepôt between Crete and the other Aegean islands and the Greek mainland.

The Minoans and pre-eruption Thera

The British archaeologist Sir Arthur Evans was searching seal-stones near Knossos, on Crete, for examples of a pictographic script, when he discovered a great pre-Hellenic civilization which originated at the end of the New Stone Age (Neolithic), about 3000 BC. He named it 'Minoan', after the legendary King Minos of Crete. Archaeologists follow him in dividing the Minoan culture into three periods: Early (now generally reckoned from 3000 to 2200 BC, the Age of Copper), Middle (2200–1550 BC, Early Bronze Age) and Late (1550–1180 BC, Late Bronze Age). These cultures developed and flourished in the Aegean area generally, with their centre on Crete, until about the middle of the second millennium BC (Evans 1921–35).

The Early Minoan culture was characterized by relatively unsophisticated village life, with some copper tools and implements, carved stone vessels and fine gold and silver jewellery.

The Middle Minoan was Crete's golden age. By this time Minoan civilization was highly advanced. Its script, Linear A, known from nowhere else, has not yet been deciphered but the simple linear signs of which it is composed suggest that it was probably derived from an older hieroglyphic picture-writing script. In this period sea trade became well established, not only with mainland Greece, but also with Anatolia, Syria and even Egypt and Libya. Cities and palaces, including the great palace at Knossos, were built. The intricately designed Knossos palace was a highly sophisticated building, with a 'modern' water system that was millennia ahead of its time, with copper piping supplying under-floor piped water, numerous bathrooms, glazed baths, flush toilets and piped sewers. Leisure activities included elaborate board games, some using dice, boxing and wrestling matches, and extraordinary, possibly religious, daring acrobatic games with bulls, in which both young men and young women participated. They evidently vaulted over the bull's back from the front, clearing the head and horns and turning a somersault before landing – clearly a dangerous game of thrills which must have ended tragically on occasions.

Three constantly recurring and unique motifs in Minoan culture and religion are the bull, the maze or labyrinth, and the *labrys* or double-headed axe. The legend of the Minotaur (i.e. the bull of Minos), a bull-headed monster waiting at the centre of the labyrinth to destroy its would-be slayers, may have arisen from the practice of holding the dangerous bull-games in central palace courtyards surrounded by a maze of small inter-connecting rooms and passages.

In spite of the danger and probable bloodshed involved in the practices with bulls, battle heroics and feats of arms were clearly not regarded by the Minoans as of great importance. Instead, peace, comfort, luxury and leisure seem to have been enjoyed to the full. Minoan palaces were not fortified. Warlike images are absent from the culture and no monumental statues have been found. A remarkably modern aesthetic appreciation was expressed in decorative arts such as ceramics, jewellery, engravings, carv-ings, miniatures, frescoes and paintings. Men are aways shown as clean-shaven and bare-chested, with a loin cloth and tight belt. Women used make-up, wore earrings and necklaces, and had a distinctive hairstyle with long ringlets at the back and often a single long ringlet falling in front of each ear. Their ceremonial dress for sacred dances consisted of a full-length, multi-tiered skirt and a short, tightly waisted open bodice that left the breasts fully exposed. Women enjoyed a high social position, had specific property rights and played a central role in religious life, providing priestesses of the Mother Godess. It has been suggested that the system of government was a matriarchy.

The Late Minoan period, like the Early period, was itself divided into three subunits, I (1550–1450 BC), II (1450–1400 BC), and III (1400–1180 BC). Although characterized by extremely fine pottery, it was the beginning of the decline of Cretan power and influence and the rise of Mycenae on the Greek mainland.

The 'Minoan' eruption

Finally, after 200 BC, renewed activity within the Bronze Age caldera formed the emergent islands of Palea Kameni and Nea Kameni. The caldera had a maximum depth of over 500 m in the north, which has now been reduced by about 100 m. (The single, oval, compound island Stronghyli was formed during the Pleistocene by the extrusion of lava and other materials, and in the past 200 000 years there have been 12 major eruptions on the island and numerous minor ones (Druitt and Francaviglia 1990).) In 18 000 BC, eruptions resulted in the formation of a collapse caldera 6 km in diameter in the south of the island, which may have been open to the sea in the south-west (Aston and Hardy 1990,

Heiken *et al.* 1990), and there may also have been a partially flooded northern caldera (Druitt and Francaviglia 1990).)

The volcano had been dormant for centuries when in the late Bronze Age (Late Minoan), some time in the second millennium BC, several decades of possibly continuous earthquake activity probably caused some of the population to leave the island. This happens today, of course; for example, earthquakes that continued for days or weeks caused evacuation of the population near the volcano El Chichon in Mexico in 1982 and at Pozzuoli near Vesuvius more recently. People returned to Stronghyli, however, and a few years later there may have been prolonged minor tremors which fortunately again gave the 30 000 or so inhabitants of Akrotiri time to pack their belongings and flee (Cadogan (1990) noted that articles of really high value were missing from the remains). Small but impressive precursory phreatic eruptions consisting of steam blasts (from underground water), and phreatomagmatic activity (involving the interaction of water with quantities of magma) may have accelerated the evacuation (Heiken and McCoy 1990). Then, probably about 2 years later (Blot 1978), with the island now uninhabited, came the cataclysmic eruption. The central volcano, 'Pre-Kameni', erupted with tremendous force.

Sparks and Wilson (1990) and Freidrich (2000) have summarized the probable sequential phases of the eruption, which almost certainly ran into one another without sharp breaks. The first, Plinian phase, which lasted for one or very few days, consisted of Plinian-type pumice eruptions (in which all eruption products are ejected in the form of tephra and gases by cataclysmic explosions) from a vent on the Pre-Kameni island only slightly above sea level, forming an eruption column which reached a height of about 37 km. Stronghyli was covered in a rain of pumice and debris which formed a layer up to 7 m thick. Currents carried floating pumice to the south, but finer pumiceous material in the column was carried to the east by prevailing winds, reaching Anatolia and the eastern shores of the Mediterranean. Very fine dust and aerosols reached the stratosphere, where jet streams carried them to the north, spreading them over the atmosphere of the northern hemisphere, reaching at least as far as Greenland.

Very violent phreatomagmatic reactions occurred when sea water reached the fluid magma in the widened feeding vent of Pre-Kameni. Clouds of fine ash suspended in steam spread in a low trajectory at some 200 km/hr from the eruption centre, filling the caldera, overspilling its walls and sweeping down the outer slopes to the sea, flowing round the highest peaks. About 2 km^3 of fine pumice was emplaced on Santorini during this phreatomagmatic, base-surge second phase, as during the first phase, very large rocks were ejected in high-energy explosions; car-sized blocks were flung up to places 10 km from the eruptive centre.

Large phreatomagmatic explosions with ignimbritic eruptions followed, as a third phase, and the poorly fluidized, low-temperature pyroclastic flows formed massive, poorly sorted deposits including fragments of non-volcanic rock from the collapsing lower regions of the volcanic edifice. These explosions were violent enough for the pyroclastic flows, as in the second phase, to carry over the 200–300-m-high caldera rim and race down the outer slopes. After this there were floods, possibly generated by repeated avalanches, and subsidence of the caldera, the floor of the northern basin, where the Pre-Kameni island had been situated, foundering to over 500 m below sea level.

Altogether some 39 km^3 of magma and rock fragments were extruded, probably in only 3 or 4 days, representing about 28–29 km^3 of dense rock (Pyle 1990a).

After the eruption, some 50–75 km^3 of pyroclastic deposits, layers of ash and lava rocks, covered the Santorini group, including the Minoan settlement at Akrotiri, to depths of from 25 to 60 m. The 'Pre-Kameni' island had disappeared, and in its place was a 500-m-deep caldera. The almost complete ring of Stronghyli was broken up into the islands of Thera and Therasia, with Aspronisi a minor island remnant, these islands now representing the shattered basal wreck of Stronghyli. A land area of 43 km^2 out of approximately 103 km^2 was lost, but about 25 km^2 was gained by emplacement of eruption deposits (Aston and Hardy 1990). The pyroclastic flows that moved over the caldera rim and down the slopes into the sea had extended Thera's eastern lowlands and incorporated the small offshore islet of Monolithos.

The steep cliffs on the inward-facing coasts of the islands are the above-water parts of the wall of the submarine, compound, tertiary caldera. The caldera's deepest, northern basin was originally 500 m deep, the southern part 200–300 m deep, but slumping of the walls has given the caldera its present form and depth, and the northern basin has been shallowed by 110 m. The present caldera is 11 km wide, 7.5 km long and 390 m deep, and has four times the area of that of Krakatau, but is rather smaller than that of Long Island. Its volume of 80 km^3 compares with Krakatau's 50 km^3, Crater Lake's 100 km^3, Long Island's 100 km^3 and Taal's 240 km^3 (Decker 1990). Since the great eruption, several islands have emerged from the centre of the caldera. They are now represented by the islands Palea Kameni and Nea Kameni, where the active vent is now sited.

The effect of the Theran eruption on Crete

The collapse of the large volume of pyroclastic material into the sea at Thera during the Minoan eruption generated a tsunami (marine pressure wave) which battered the Minoan coastal settlements on Crete, 100 km to the south (Fig. 6.5). Yokoyama (1978) calculated that the tsunami probably raised sea level by about 60 m at its point of origin and took some 25 minutes to reach Herakleion (near Knossos) on Crete, by which time the level had fallen to 11 m above normal. Damage from tephra fall from the ash cloud of the volcanic column was limited. Although it certainly affected eastern Crete, Thorarinsson (1978) believed it

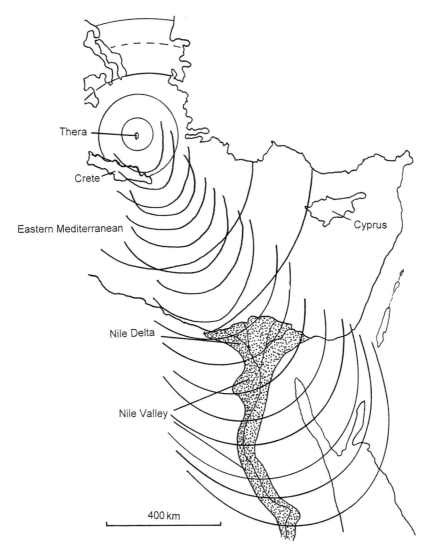

Figure 6.5 Santorini: the wider effects, as illustrated by extension through Crete to the Nile Delta and beyond from the volcanic epicentre at Thera. (After van Bemmelen 1971.)

unlikely that either tsunamis or eruptive fallout played a direct, decisive role in the fall of the Minoan empire. Ash fall and evidence of tsunami damage in the ports of Crete is slight, and he believed that the destruction of the Cretan palaces was two centuries too late (in the middle of the fifteenth century BC) for the eruption of Thera to have been the direct cause.

From archaeomagnetic studies, Downey and Tarling (1984) concluded that the Thera eruption had two phases which were contemporaneous with two phases of destruction on Crete, the first corresponding to devastation in central Crete, the second, from 10 to 30 years later, affecting eastern Crete. Manning (1989,

1990b) concluded that, although the archaeomagnetic evidence was not as clear as was first thought, radiocarbon dates of short-lived material from both Crete and the Akrotiri excavations on Thera, and the stylistic similarity between frescoes on Thera and Crete, support Downey and Tarling's conclusion and suggested that central Crete was devastated first and eastern Crete some years afterwards.

The second bout of devastation on Crete, involving sites at which only Late Minoan IB pottery was found, was the more destructive and long-lasting, and there was evidence of widespread fires. Cores from the eastern Mediterranean sea bed, however, revealed that central Crete would not have received sufficient ash fall to have caused the burning and destruction that is shown in the archaeological record (Watkins *et al.* 1978), and although eastern Crete received more fallout than central Crete, most of the Thera ash fell to the east of Crete in the eastern Mediterranean and on south-west Turkey (Pyle 1990a). This major destruction on Crete is thought by some (e.g. P. Faure, in discussion of Thorarinsson 1971) to have been the result of a war of conquest by the Mycenaeans from mainland Greece who took over the Minoan state, rather than the effects of the Thera eruption.

Although evidence from frost rings in trees, Greenland ice cores and radiocarbon dating (see box) suggests that the eruption occurred in the 1620s BC, the preponderant archaeological interpretation of the ceramic data places the Cretan destruction almost two centuries later than this (about 1450 BC). Evidence from a record of an unusual meteorological event in Egypt (see below), if caused by the eruption, would place it between the two, at 1550/39 and 1528/17 BC.

Radiocarbon dating

The radiocarbon dating of organic material found in archaeological remains has been used increasingly in recent decades. This dating method depends on the fact that when cosmic rays reach the atmosphere a radioactive isotope of carbon, ^{14}C or carbon-14, is formed as neutrons are added to atoms of nitrogen. Thus a proportion of the carbon in the atmosphere is always of this isotope. Plants absorb the radioactive carbon as well as the stable isotope, as they take in carbon dioxide, and this is passed to the animals that feed on them. When the plant or animal dies, however, the radioactive carbon decays at a constant rate, such that after 5730 years it has been reduced by half. Thus if the measured amount of carbon-14 in a dead organism is compared with the original level of that isotope in the atmosphere, one can calculate how long the organism has been dead, and thus the age of any material associated with it.

Samples used for dating should not be long-lived material that was already old at the time of death. The outer annual rings of a tree trunk may be several thousand years younger than the innermost rings. Short-lived materials such as seeds harvested in the year of production, or thin branches with only a few annual rings, are preferable.

A calibration curve is necessary to convert the carbon-14 age (BP, in effect meaning before 1950) into an absolute age (BC). For the middle of the second millennium BC, for example, the radiocarbon age may be some 300 years more than the calendar age. However, because the strength of cosmic rays is not constant, the amount of carbon-14 in the atmosphere varies over time, and the calibration curve takes into account natural changes in the atmospheric radiocarbon levels. It is thus not a straight line, but a 'wiggly' one. Moreover, in the sixteenth and seventeenth centuries BC, 'ambient' levels of radioactive carbon do not provide a monotonic curve, and radiocarbon levels of later seventeenth century are almost identical to those of the mid-sixteenth century. This means that over this period radiocarbon dates of archaeological material may be matched with not one but two dates when corrected in this way (Housley *et al.* 1990). A method of 'wiggle-matching' (matching the wiggly shape of part of the atmospheric radiocarbon curve to a segment of the curve for the organic archaeological material) has been devised to obviate this problem.

Hardy and Renfrew (1990), reviewing dates indicated by Akrotiri artefacts, as well as those from ice core and tree-ring correlations (see below), tended to accept a considerably earlier date of about 1626 to 1648 BC for the Thera eruption. They believed, however, that the archaeological record of the Minoans' demise was unlikely to be out by well over a century, and therefore concluded that the destruction of Minoan Crete was later than the Thera eruption and must have had some other cause. As noted above, this belief is supported by evidence that, although tephra fall certainly damaged eastern Crete significantly, most of the Thera ash fell to the east of Crete (Watkins *et al.* 1978, Sullivan 1988, Johnstone 1997).

Considerable debate about the eruption's role in the fall of the Minoan empire continues, with opinion now tending to support Thorarinsson's (1978) conclusion, that it is unlikely that either eruptive fallout or the associated tsunami played a direct, decisive role in the fall of the Minoan civilization on Crete. In summarizing the results of the latest Thera conference, Hardy and Renfrew (1990) concluded that although the eruption's environmental effects may have been so extensive as to have undermined the political stability of the Minoan state and triggered its decline, the Thera event was too early to have been involved directly in the physical destruction of the Minoan palaces and the downfall of the Cretan civilization.

Surprisingly, this massive explosive eruption on Santorini does not seem to have been recorded, either by the Cretan Minoans or the ancient Egyptians.

The effects can be deduced only from geological and archaeological records, which provide conflicting indications.

The earliest eruption that can be dated exactly is probably that of Vesuvius in AD 79 (Zielinski and Germani 1998a). Earlier eruptions must be dated by a number of detection methods. Traditional archaeological evidence is the attribution of pottery remains on stylistic grounds to a particular period, where possible involving reciprocal matching with another, better-dated culture. Thus, for example, Minoan pottery found in Egyptian excavations at a particular known period of Egyptian history, together with Egyption material from that period found on Crete together with the same style of Minoan pottery, would be good archaeological evidence for the dating of the Cretan pottery style.

Recent studies in dendrochronology (see box) have identified growth-ring anomalies at around 1627 BC in very long-lived bristlecone pines in western United States, Irish bog oaks and Anatolian juniper wood. Ice cores in Greenland have revealed high-sulphate peaks in layers dated at about the same time (see box). Clearly there was some large atmospheric perturbation at that time, characterized by high sulphate levels in the atmosphere and an altered plant growth environment over a wide area of the northern hemisphere. And Santorini's eruption was a prime suspect. The date $c.1627$ BC was cited almost as though it was accepted as the definitive date for the eruption, although one or two scholars urged caution and noted that the culprit might be pinned down decisively, one way or another, by the finding of volcanic material in the ice-core layer at this date that had the high-sulphate peak, and comparing it with material from Santorini at this annual layer, analysing it chemically and physically (for oxide spectrum and refractive index) and comparing it with samples from Santorini itself. Such material was found, in the form of volcanic glass shards, in the GISP2 (summit) ice core in Greenland, in the same layer of ice that had the large sulphate spike. In this core, the layer that corresponded to 1623 BC was thought to be almost certainly the same 'spike' as found in other cores at around this date. The shards were analysed chemically and compared to similar material from the co-ignimbrite deposits on Santorini. The answer was in the negative. The shards could not have come from Santorini. This conclusion was questioned (Manning 1998) on the grounds that the wrong Santorini material had been used as comparison – if the Plinian material of the early eruption phase was used, rather than material from the later, co-ignimbrite phase, there was good correspondence with the ice-core layer material.

Dating by tree-rings

More recently, another method of dating eruptions has been developed. This uses a record of past climate, and is only indirectly linked to eruptions through the climatic perturbations that they are known to induce. The record is in tree-rings, and the method was first developed by examining the rings of exceptionally long-lived trees, such as the bristlecone pines (*Pinus aristata*) of California, which may be up to about 4000 years old and provide a tree-ring record to which radiocarbon ages can be calibrated. In the same area as the living pines, remains of trees that died more than 4000 years ago have been analysed, and a continuous record of annual rings has thus been established that goes back some 9000 years.

It is now well established that the emissions of an explosive eruption, the fine dust and sulphate aerosols, which are carried for weeks, months or years in the atmosphere, affect the Earth's climate by reducing the amount of solar radiation reaching the Earth's surface. This may be reflected in the growth of trees, which may suffer frost damage, reduced growth or, in certain circumstances, abnormally increased growth, and these growth anomalies can be recognized in the tree-rings. In the bristlecone pines, abnormal frosts in certain years show up as characteristic anomalous growth rings, where the cells in the wood had been killed, which are recognizable as 'frost rings'. Bog oaks in Ireland and oaks and pines of the Danube in Germany were also found to provide a record by a characteristic series of close-set rings, indicating markedly reduced growth over a period. In the Anatolia region of Turkey (in the direct path of the eruption column from the Santorini eruption) juniper trees similarly record climatic changes in their rings, but in this case the anomaly is one of rings being laid down further apart than usual, recording abnormally increased growth. As noted above, a direct link to any particular eruption (as is also the case for ice-core sulphate concentrations) or even to volcanism in general, is lacking. The tree-ring method has the drawback that it does not relate directly to volcanic eruptions. It signals changes in the environment that have affected tree growth, but there is no connection other than coincidence to link the signal with volcanic eruptions in general (for many factors other than volcanism may be responsible for climatic anomalies), let alone any particular eruption. It has the advantage that it does provides precise dating.

Ice-core dating

The sulphur-rich gases, including sulphur dioxide, emitted into the stratosphere by explosive volcanic eruptions combine with water to produce sulphuric acid aerosols that drift over vast distances in jet streams and remain in the atmosphere for weeks, months or years. Normal precipitation of rain or snow eventually scavenges the acid (as acid rain or snow) and deposits it on the Earth's surface. In polar areas it is deposited as snow, and the precipitation of thousands of years is stored as stacked compressed layers of snow (ice) in the ice sheets, which in the Greenland Ice Shield, for example, are about 3 km thick. Each ice layer has chemical and optical properties that are peculiar to the year in which it was deposited, so that the layers can be counted, just as one can count annual growth rings of trees, and the age of any particular layer can be determined.

If the ice sheet is drilled, a long core of compacted annual ice layers can be obtained, providing a record of atmospheric conditions going back some 200 000 years. Minute amounts of acid in layers can be detected by measuring the electrical conductivity of the ice (the acid increases the conductivity) and sulphuric acid can then be identified and determined by chemical methods. The sulphuric acid signal, and the eruption or eruptions that produced it, can thus be dated. There is thus a direct record of explosive volcanic activity over a long period of time, stored in the polar ice sheets.

In this way sulphuric acid ice-core signals (sharp 'spikes' of sulphuric acid concentration) have been correlated with the Indonesian explosive eruptions of Krakatau (1883) and Tambora (1815). Large eruptions close to the equator are recognizable in ice layers of both hemispheres, but some eruptions at high northern latitudes (e.g. Katmai, Alaska in 1912) left only a weak signal in drill cores from southern Greenland because of the northerly course of stratospheric wind in the northern hemisphere.

However, although a prominent ice-core sulphate spike is direct evidence of explosive volcanic activity at that date, identifying the eruption responsible is often difficult, because factors such as proximity and bearing relative to air currents, relative proportion of sulphur emissions to total emissions, and possibility of multiple eruptions, may confound the identification. The crucial evidence that is needed to tie the spike to a particular eruption is the presence of fine tephra products of the eruption, such as shards of volcanic glass, in the ice layer. Relative concentrations of elements and oxides in such material provide a 'spectrum' that is typical, a characteristic fingerprint, of each eruption. The fragmentary material in the core may be analysed physically and chemically (e.g. refractive index and oxide spectrum), and compared, for corresponding characteristics, with similar deposits at the site of eruptions suspected of being responsible. This has been done for a number of eruptions (e.g. Zielinski *et al.* 1996). Even when this is possible, however, there is a complicating factor: characteristics of the eruptive products of an explosive volcano do not remain constant throughout all phases of the eruption (Manning 1998, Zielinski and Germani 1998b).

In the late 1990s the evidence from tree-rings and sulphate concentrations in deep ice cores appeared to support an 'early' date for the eruption, in the seventeenth century (around 1620 BC), while archaeological evidence generally pointed to an eruption late in the sixteenth century BC.

The most recently proposed date for the Thera eruption is about 1650 BC (Manning *et al.* 2001). This is based on unusually high sulphate concentrations found in annual layers of compacted ice in deep Greenland ice cores (Hammer *et al.* 1987, Clausen *et al.* 1997), and the radiocarbon dating of tree-rings in southern Turkey, which lies directly in the path of the Santorini eruption cloud. The tree-rings dated in Turkey were from timbers used to construct the grave chamber (perhaps of the legendary King Midas or an ancestor), the world's oldest standing wooden building, in the 'Midas' mound tumulus at Gordion, Anatolia. Unusually widely separated rings, representing unusual growth, were found over a period of 3–5 years in juniper, cedar and pine timbers. The date of the beginning of this growth anomaly is around 1650 BC. In the Greenland ice cores, an unusual 'spike' of sulphate concentration occurred in the year 1645 ± 7 BC. Past concentrations of atmospheric sulphuric acid result from explosive volcanism; the acid is scoured from the atmosphere by rain (acid rain) and snow, and each years's snowfall is compacted, layer on layer, on the great polar ice sheets. A long-running, dated record is thus preserved of the characteristics of the atmosphere through which each year's snowfall. It has been shown that this signal did not originate in the high northern hemisphere (for example, Iceland) and no southern hemisphere eruption is known from this period. It was suggested that Santorini's eruption cloud was responsible for the sulphate spike, and the analysis of associated glass shards appears to confirm this.

The Great Date Debate: when did the 'Minoan' eruption occur?

The Santorini eruption was the greatest volcanic cataclysm of the second millennium BC. Its connection with historical events such as the demise of the Minoan civilization on Crete (see above), the 'Aten heresy' in ancient Egypt, the biblical Exodus of the Hebrews from Egypt, the fall of Jericho, or the end of the Xia Dynasty in China, depends crucially on precise dating of both the eruption and the events in question.

In 1979, at the second Thera conference, a proposed link between the eruption and the Cretan destruction was largely dismissed. Researchers then placed the date of the eruption at around 1500 BC (the so-called 'low' or 'late' chronology), the date originally proposed by Marinatos (1939) and supported more recently by Warren (1984), for example. This was at least 50 years earlier than the horizon of destruction evident on Minoan Crete. Thus the eruption, and the end of Late Minoan IA, were set at no earlier than about 1500 BC, with the great destruction of Late Minoan IB on Crete at about 1450 BC. This low chronology rests on synchronisms established between Egypt and Crete by archaeologists, mainly on the basis of complementary finds of Egyptian and Cretan objects. The 1500 BC date appeared to be supported by uncalibrated results of the first radiocarbon-dated series of samples of carbonized material from Santorini. Radiocarbon results calibrated against tree-rings of bristlecone pine in California, however, gave a very different date – about 1625 BC.

In 1980, workers in Denmark published the results of studies on Greenland ice cores (Hammer *et al.* 1980). So far, three boreholes have been drilled in the ice, providing three cores, one from the north, one from the south, and, in 1992, one through the deepest part of the ice sheet, in central Greenland. The Danish workers identified a peak of volcanic activity at within 50 years of 1390 BC as corresponding to the Thera eruption. This brought the eruption date forward to a time that was now reasonably consistent with the then accepted archaeological date of the Minoan destruction on Crete (about 1450 BC).

Seven years later, however, the Danish group retracted this proposed date. It was discovered that the ice core in question had been broken and a date wrongly assigned, and a new core provided a much earlier date for the eruption – 1645 ± 7 BC. On this evidence the eruption was now again 'too early' to be linked with the Cretan destruction. The high acid signal was associated with Santorini as the only eruption of appropriate magnitude within several centuries either side of the signal. For an absolute link with this particular eruption, traces of glass that 'fingerprinted' the Santorini event would need to be found at the appropriate level within the core. Such traces were found from the El Chichon eruption in 1983, and in time it is possible that Santorini particles will be found in cores (Anonymous 1990).

Lamb (1970) had proposed that the atmospheric pollution of major eruptions (what he referred to as the 'dust veil', now known also to include fine-particle sulphur dioxide or sulphuric acid aerosols: Rampino and Self 1982, 1984) can so filter out sunlight that world temperatures are lowered. La Marche and Hirschboeck, of the University of Arizona, published a major paper in 1984 proposing that in eastern California unusual cooling caused by volcanic eruptions caused frost damage in bristlecone pine trees at 4000 m altitude, above the tree-line (but not in those several thousand feet lower down). The damage in the high-altitude pines is recognizable in the growth rings of the wood. Years in which there was severely retarded growth or frost damage show up as 'frost rings', which often, but not always, correspond with the dates of major volcanic eruptions. There was no such signal in the tree–rings around 1500 BC, nor for a hundred years either side of that date. However, in at least half of the trees sampled there was a clearly recognizable frost ring corresponding to 1627 BC, evidence of a volcanic explosion several times larger than Krakatau having taken place a year earlier. This was the only noticeable major frost event between 2035 and 206 BC. La Marche and Hirschboeck (1984) proposed that the Santorini eruption was the most likely cause. If this were so, the eruption would have taken place in 1628 BC. A 'high chronology' now had to be considered.

Support came from Europe. Samples of oak logs preserved in Irish peat bogs were part of an absolutely dated chronology spanning 7272 years, established by Baillie's group at the Queen's University, Belfast. After seeing La Marche and Hirchboeck's paper, they checked their samples and found that in six trees from four different bogs the narrowest rings in the trees' lives were produced in the 1620s BC, starting at about 1628 (Baillie and Munro 1988, Baillie 1990). In this case the signal is a band of very narrow rings for about a decade, formed, it was suggested, as a result of the effect of climatic deterioration (cold and possible flooding of these marginal environments) following the Santorini eruption, although it must be noted that Sear et al. (1987) had estimated that the aerosols would persist in the atmosphere for a maximum of 3 years. The Santorini eruption was shown to have been large enough to leave substantial amounts of ash in western Turkey (Sullivan 1988) and therefore could conceivably have affected the growth of trees in the northern hemisphere generally. Since then, investigation of bog oaks at two sites in England (in East Anglia and at Croston Moss, Lancashire), Germany and southern Ireland (Baillie 1990) have shown that 1628 was an outstandingly bad year not only for California bristlecone pines, but also for oaks in many parts of Europe.

Both tree-ring and ice-core data showed major events at dates scattered around 207, 1159 and 1628 BC, suggesting that the 1645 signal from the ice core and that of 1628 from tree-rings relate to the same event.

Warren, however, cautioned that correlation between the tree-ring data and that from high-sulphate acidity ice-core peaks is poor. Between 3435 BC and AD 1965, of 23 eruptions producing high acidity peaks, only ten correlate with American frost-ring data, and 26 of 36 frost rings have no corresponding acidity peak (Anonymous 1990). Of course, as pointed out by British archaeologist Buckland and his colleagues, climatically induced frost damage or retarded growth in trees can have a number of natural causes, other than volcanic eruptions. This is in contrast to acidity peaks in ice cores, where, if the peak is a sulphuric acid peak, one can be sure that it was caused by a volcanic eruption. Whereas the tree-ring data are circumstantial, in that they rest on a relationship between eruptions and world climate, the sulphate acidity peaks have been largely caused directly by volcanic eruptions. Archaeologists (Warren 1984, Manning 1988, Buckland et al. 1997) argued first that not all frost rings have corresponding eruptions nor all eruptions frost events, a point explicitly acknowledged from the start by La Marche and Hirschboek, and second, that there appears to be no direct correlation between the scale or geochemical character of eruptions and ice-core signals. Pyle (1990b) noted that the Santorini eruption, like that of Krakatau, was probably relatively low in sulphur, providing a correspondingly slight signal in the ice, and he believed that direct evidence for the eruption having taken place in the 1620s decade was still lacking. Zielinski *et al.* (1994) suggested that if the Santorini eruption had a low ratio of sulphur to total erupted volume, one of the smaller sulphuric acidity peaks in the 1600s BC may be its signal; they nevertheless related a high peak in 1623 BC to the eruption.

Many more radiocarbon dates of short-lived material from the volcanic destruction level at Santorini have been determined by several laboratories around the world. Manning (1990a) noted that radiocarbon dates have wide error bands and necessitate complex non-linear calibration, and that the contextual and measurement quality of many of the samples has been questioned, with the conclusion that the results of many analyses are less than satisfactory. Nevertheless, he analysed the results carefully, and concluded (Manning 1990a, 1990b) that their undeniable trend was towards a date in the seventeenth century BC, and that a date in the mid-sixteenth century BC, coinciding with Warren's date, was unlikely.

Thus the 'indirect' evidence, from records of climatic and atmospheric effects, which is fairly precise, supports a high chronology (late seventeenth century or the 1620s BC), while the archaeological or 'direct' evidence, which is less precise, tends to support a low one (late sixteenth century or around 1500 BC). The notable exception is that radiocarbon dating tends to support the other 'scientific' evidence, indicating a high chronology. Summarizing the 1989 Thera conference, Hardy and Renfrew (1990) reviewed dates indicated by Akrotiri artefacts as well as those from ice-core and tree-ring correlations, and tended to accept an earlier date of about 1626 to 1648 BC for the Thera eruption. Manning has suggested that the existing archaeological evidence is sparse and somewhat imprecise, both as to the dates of the contexts of finds and to their stylistic attribution; he noted that 'the Egyptian chronology, particularly the Egyptian relative ceramic chronology, is less precise than has been often assumed by Aegean scholars' (Manning 1990b: 92). He believed that the existing evidence from archaeology does not *require* the traditional chronology; it can plausibly fit the early chronology.

There have been five notable developments since the 1989 Thera conference, four adding weight to the high chronology case, one throwing doubt upon it. First, in 1992 a third core was drilled in the thickest layer of the Greenland ice cap, at the central 'summit'. This was over 3 km long, the longest continuous ice core drilled in the northern hemisphere. From this core, Zielinski *et al.* (1994) identified a high sulphate ion acidity peak at 1623 BC, which they tentatively related to Santorini's Minoan eruption. Later, Clausen and colleagues (1997) provided what they considered to be a more reliable date for this peak – 1636 ± 7 years BC, which corresponded closely to the date for a similar signal found earlier in the southern Greenland core. The mean of the two dates is 1640 BC.

Further important supporting evidence for the 'early' or 'high' chronology, came from tree-rings from material directly in the path of fallout from the Santorini eruption. A 1503-year tree-ring sequence in Anatolian archaeological juniper wood was examined by Kuniholm *et al.* (1996; see also Kuniholm 1990). The sequence was based on timbers used in the construction of an Old Assyrian trading station at Kültepe and (for the later part of the time range) from the 718 BC 'Midas' mound tumulus about 100 km south-west of Ankara. A remarkable growth anomaly was found in the seventeenth century BC, with tree-rings from three to seven times further apart than normal. This was correlated both with major growth anomalies at 1628 BC in the securely dated tree-ring sequences in Europe and the USA, and with the evidence from Greenland ice cores. Kuniholm's group explained the *increased* rather than retarded growth in their material by the 'unusually high and sustained moisture content and a sharp reduction in midsummer evapotranspiration'. Renfrew (1996), however, did not accept that this correlation necessarily shows the anomaly to be the result of a reduction in solar radiation received by the junipers, and that a fuller explanation is needed. But, in all cases, in the Turkish, European and American tree-ring sequences, the anomaly stood out as the only one in the eighteenth to fifteenth centuries BC. However, a second matching anomaly was found between the Anatolian and European tree-ring data sets at 1159 BC, again the only important growth anomaly in either chronology for several centuries, and the match again coincided with ice-core data from Greenland, in this case indicating the eruption of Hekla 3 in Iceland. The discovery of this second matching anomaly strengthened the case for the first one, and thus the significance of the Anatolian evidence.

The one development that tends to throw doubt on the high chronology theory is further questioning of the correlation of climatic and atmospheric pollution signals with the Thera eruption, rather than with some other event. The possibility has been raised that Yiali, another island volcano at the eastern end of the Aegean arc, also erupted at about this time, and may have been involved (Galloway and Liritzis 1992), and at least three other volcanoes (Vesuvius, Mount St Helens, Aniakchak II in Alaska) erupted between 1650 and 1550 BC (Buckland *et al.* 1997). Renfrew (1996) noted that, because the Minoan eruption seemed to be securely linked to the beginning of Egypt's Eighteenth Dynasty by finds of Santorini pumice in Lower Egypt, to adopt a date in the seventeenth century BC for the Minoan eruption would imply very substantial changes (over a century) in the historical chronology of ancient Egypt, changes that would have enormous implications for the archaeology and prehistory of Egypt and the Old World generally. He believed, however, that such an important revision would need to be based on more than the existing evidence. A particle of identifiable Thera tephra in an ice core at the appropriate level would tie the signal to Thera, but this has not yet been found.

So, from this considerable debate and a large body of research and scholarship in various disciplines, what is the lay person to conclude about the date of the Thera eruption and the end of Late Minoan IB? And, enjoyable though the search for the correct date may be, is it important?

Although the matter is still unsettled (Harding 1989, Manning 1988, 1989, 1990a, 1990b, Renfrew 1990a, 1990b, 1996), and a mid-sixteenth-century date is still an unlikely possibility, the evidence for an early chronology appears to be growing, and it seems likely that researchers will settle on a date in the second half of the seventeenth century BC, in the 1620s, over a century earlier than the traditional date, at least as a working hypothesis. Keenan (2003) has suggested recently that volcanic ash from the Greenland ice core is indeed not from Thera: the dating question that has provided such heated debate (and spawned a vast literature) over recent decades seems set to continue for some time to come.

Whatever the outcome of studies on the role of Thera's eruption on the decline of the Minoan civilization or on its effects in more distant lands, Fisher and his co-authors (1997) have reminded us of Schliemann's 1880 and Nilsson's 1930s proposals, that natural events such as this provided Bronze Age people with the stuff on which Greek mythology is based.

Biological recolonization of Thera

Today, Santorini is one of the most arid of Greek island groups, and lacks trees, such as *Juniperus phoenicea*, oaks like prickly-oak, *Quercus coccifera*, and the drought-resistant wild pear, *Pyrus amygdaliformis*, that are typical of maquis (Rackham 1990). The only true shrub of maquis vegetation, the highly drought-resistant *Pistacia lentiscus*, is present on Santorini, but rare.

Recolonization of the devastated Thera by plants and animals occurred too long ago to be monitored, but Heldreich (1899, 1902) noted that in 1881 many plant species that were common on the less-affected islands of the Cyclades group were absent from Thera. This evidence of floral impoverishment was supported more recently by Hansen (1971) and Diapoulis (1971), and Frör and Beutler (1970) concluded that the herpetofauna (reptiles and amphibia) were also relatively poor. Even when 500 species of plants had been found on Santorini, Hansen (1971) attributed the 'scarcity of species' to the porous soil and general lack of free water and ground water. He noted that there are almost no endemic species and regarded the flora as a new one, assembled since the Minoan eruption. These biologists believed that the present biotic impoverishment of the island is good evidence of total extirpation of the biota 3500 years ago.

Their view was not universally supported, however. Rackham (1978) was of the view that the flora is not specially impoverished. He noted that the whole southern half of Stronghyli was not involved in the Minoan eruption but was smothered in a deep layer of ash fall that in most areas would have destroyed the vegetation. He thought it possible, however, that some plants could have survived on the cliffs of the limestone mountains in the east, and cited the date palm, which may have grown wild in the north of the island and is extremely fire-resistant, one of the most likely species to have survived (Rackham 1990). Schmalfuss and Schawaller (1984) did not regard the flora of 515 vascular plants as being relatively impoverished, nor did they find convincing evidence that either the herpetofauna or the tenebrionid beetle fauna (Grimm 1981) was depleted compared to those of unaffected islands. In direct contrast to the earlier workers, they concluded that extermination of the biota may not have been total, or that if it was, all possible niches had been filled by colonists in the subsequent 3500 years. Raus (1991), however, found the present species number was not low compared to islands of similar size that had not been devastated by eruption, and believed that the island is still unsaturated.

There is thus serious disagreement on the interpretation of data with respect to the recolonization of Santorini by plants.

CHAPTER SEVEN

Long and Ritter Islands, Bismarck Sea

Some time around the middle of the seventeenth century AD there was a volcanic eruption at the western margin of the Pacific which, in terms of energetics, was several times larger than the Mount St Helens eruption and about the size of that of Krakatau (Blong 1982), yet the Western world has no record of the event. This was the catastrophic eruption of Long Island, some 50 km north of New Guinea and about 150 km east-north-east of Madang, in the Bismarck Sea of the western Pacific.

There is no written record of this event, but accounts of the eruption's effects have been passed down from generation to generation by the tribes of the region in oral legends and traditions and in this way have been preserved among more than 30 different language groups (Blong 1982). Because of the effects of the extensive ash clouds, people spoke of the eruption as 'The Time of Darkness'.

The Bismarck volcanic arc

The Bismarck Sea lies north of New Guinea and New Britain, and along its southern margin runs a 1000-km-long chain of volcanoes, the Bismarck volcanic arc. The arc comprises a line of island volcanoes north of the New Guinea coast (Fig. 7.1) and a line of volcanoes along the length of New Britain. It extends from Vokeo, about 100 km north of the Sepik coastal town of Wewak in the west to the caldera port of Rabaul at the eastern end of New Britain. Johnson (1976) showed that Long Island is part of a distinctive western section of this arc that extends as far east as, and includes, the volcanoes of the western cape of New Britain, Cape Gloucester.

The eastern Bismarck arc, which includes all the New Britain volcanoes except those of its western cape, is associated with underthrusting of the 40-km-thick floor of the Bismarck Sea (on the South Bismarck tectonic plate) by the denser, 10–15-km-thick floor of the Solomon Sea, which is being consumed by subduction along the 8000-m-deep New Britain trench. Although it has several unusual features, the eastern arc nevertheless exhibits characteristics of the region's classical island arcs generally: an arcuate form, an associated deep submarine trench, volcanoes built up over the deeper parts of a

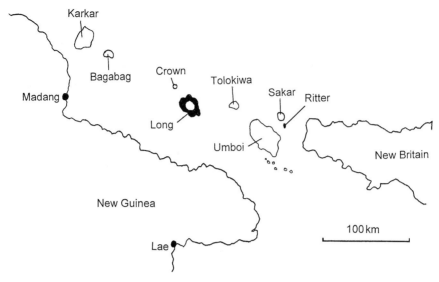

Figure 7.1 Outline of the Bismarck volcanic arc, to indicate position of Long Island and Ritter Island.

northward-dipping Benioff zone, and a seismic zone of earthquake foci up to 565 km deep.

The western arc, in contrast, does not appear to have any associated trench nor any clear seismic zone, and does not comprise a simple island arc. Its lavas differ in composition from those of the eastern arc (Johnson *et al.* 1972). Johnson (1976) speculated that in the Upper Eocene and Oligocene/Miocene a northward-moving Indo-Australian plate, with New Guinea near its leading edge, was converging with an Adelbert–Huon island arc. Northward-moving oceanic crust was subducted beneath the arc, and the 'Finisterre volcanics' represent island arc magmas related to the descending part of this northbound slab. In the Lower Miocene, New Guinea itself finally collided with the island arc, and as plate convergence continued the arc became welded to New Guinea's northern edge as the present northern coastal ranges (the Adelbert, Finisterre and Huon peninsula mountains). The convergence resulted in the continental land mass of New Guinea proper being foreshortened, warped and uplifted at this time, as the New Guinea Highlands were formed. Descent of the subducting continental slab beneath the Bismarck plate continued and steepened, resulting in the Quaternary volcanism of the western Bismarck arc, of which Long Island is a part.

The western Bismarck arc thus comprises Vokeo, Kadovar and Bam, offset about 25 km north-eastwards from a more easterly sector of the arc: Manam, Karkar, Bagabag, Crown, Long, Tolokiwa, Umboi, Sakar and Ritter islands and a number of volcanoes on Cape Gloucester. Ten of these western arc volcanoes have erupted in the last century, and five, Manam, Karkar, Long, Ritter and

Langila, which lie almost adjacent to one another along 420 km of the arc, erupted within a period of 8 months in 1974 (Cooke *et al.* 1976). Manam is now the most active of the island volcanoes; pyroclastic flows (*nuées ardentes* or glowing clouds) appear to have featured regularly in its frequent eruptions in historical times (Palfreyman and Cooke 1976). The eruptions of Karkar, Long and Ritter were after periods of repose lasting for over eight decades (McKee *et al.* 1976).

Mennis (1978, 1981) noted that Austronesian-speaking peoples from several small islands and mainland localities in the Madang area and along the Rai coast to the east of Madang all trace their ancestry to an island called Yomba, which exploded between eight and ten generations ago. They all have traditional stories of once being together on that island. Linguistic studies support a common origin for these peoples; the languages of the Madang area and those of the Rai coast belong to related subfamilies of the Austronesian language group (Z'graggen 1975). The people interviewed by Mennis all believed that Yomba blew up and disappeared before the eruption of Arop (Long Island) which these seafaring people knew well, and about which many of them gave details. It appears, then, that the stories do not concern Long Island's eruption. Although accepting that there was no geological evidence for his proposal, Mennis (1981) raised the possibility that another volcanic island, situated between Bagabag and Crown islands (Fig. 7.1) (Ball 1977), exploded in the seventeenth century, leaving as its remnant only Hankow Reef, which has a submarine base of considerable area. Long Island, Crown Island and Hankow Reef form a line intersecting the general trend of the western arc at about 30°, and immediately to the south of Long Island, earthquakes with foci at intermediate depths (150–230 km) have been frequent in the last 20 years, defining an almost vertical zone of activity.

Long Island's eruptive history and the last eruption

The British seaman and buccaneer William Dampier named Long Island, in 1700, after its elongate north–south shape as seen in profile from the western approach, a 'long Island, with a high Hill at each End'. The island is called Pono by its present inhabitants and Arop by people on the New Guinea mainland. It is actually shaped like a roughly hexagonal doughnut with a maximum diameter of 27 km and an overall area of about 425 km^2. Its chief feature is a huge central caldera, the centre of the doughnut. Long Island's main volcano, now represented by the caldera, was of andesite, and situated between two smaller, older, eroded basalt stratovolcanoes (Fig. 7.2). These, Mt Réaumur (1280 m) in the north and Cerisy Peak (1112 m) in the south, as in Dampier's day, punctuate, like inverted commas, the island's long, low profile.

Oral tradition suggests that the island was once higher than its present high point (1280 m) or that of adjacent Tolokiwa Island (1400 m), indicating that

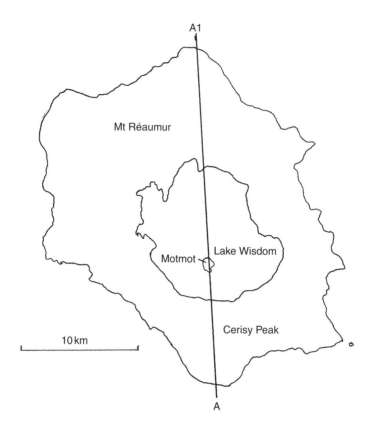

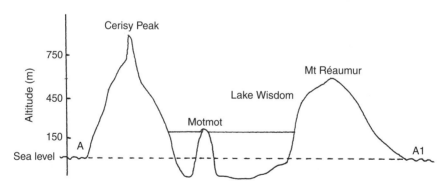

Figure 7.2 Long Island and Motmot. Upper: outline of Long Island, with Lake Wisdom and Motmot; lower: profile of transect A to A1 indicated in upper figure. (After Ball and Johnson 1976.)

a high summit may have been lost in the explosive eruption (Blong 1982). Dampier, who described the island as forested, recorded and sketched the island's northern and southern mountains but made no mention of a central mountain or mountains, and showed none in his sketch, which matches well the profile of today's island. The weight of evidence suggests that Long's eruption occurred well before Dampier's visit, around the middle of the seventeenth century.

Long's eruption was at least of Krakatau magnitude; its explosive energy was about 10^{26} ergs and more than 30 km^3 of island was lost. The volcanic explosivity index (Newhall and Self 1982) is now assessed at 6 on an 8-point scale, the same as Krakatau (1883), with only Tambora, Indonesia (1815) and Baitoushan, China (1010) being rated at 7 in the Christian era (Polach 1981, Zielinski *et al.* 1994).

The great caldera, which was formed by a series of collapses associated with pyroclastic flows (see below), with an area of about 120 km^2 and a maximum diameter of about 13.5 km, is slightly larger than that of Oregon's famous Crater Lake volcano. As with the Oregon volcano, a freshwater lake accumulated within Long Island's caldera. Lake Wisdom (see Chapter 9) is surrounded by the rough doughnut ring of Long Island, which itself has a land area of about 330 km^2. The lake surface is some 190 m above sea level and the lake reaches depths of more than 360 m, just over half the depth of Crater Lake (604 m). Long Island's caldera walls are cliffs rising from 200 to 300 m above the lake surface and, except for two embayment areas on the western and eastern shores, they run steeply into the lake without a shallow shore zone. In the north the caldera rim reaches a height of 450 m above sea level and the caldera thus has a maximum depth of over 620 m.

Ball and Johnson (1976) and Pain *et al.* (1981) have provided an excellent volcanic history of Long Island and the lake island of Motmot (Fig. 7.3). Because there is very little evidence of lava flows in the caldera's steep walls, it is thought that before the cataclysmic caldera-forming event Long Island consisted largely of the aerial component of a pyroclastic volcano or volcano complex with a submarine base of from 60 to 70 km in diameter. In the eruptive history of Long's central volcano, Pain and his colleagues recognize at least four periods (Pain *et al.* 1981).

Some time over 16 millennia ago, an initial series of eruptions, with quiet intervals, deposited a 'basal unit' layer of bedded, sorted, air-fall tephra reaching a thickness of over 4 m, resting on lava. About 16 000 years ago this was followed by three catastrophic eruptive events, each with pyroclastic flows. The events were separated by relatively quiet intervals of thousands of years, and each consisted of three phases.

First was a Plinian phase, when a layer of unsorted, unbedded, air-fall pumice lapilli (a form of volcanic ash), accretionary lapilli and lava fragments in a light-coloured earthy matrix was deposited. This developed into a Peléan phase, with pyroclastic flows arising from the base of a collapsing high eruption column and flowing down the volcano's flanks, circumventing obstacles and thus leaving no deposits on the high ground of the older volcanoes, for example. On the slopes and lowlands these flows resulted, in each event, in a thick

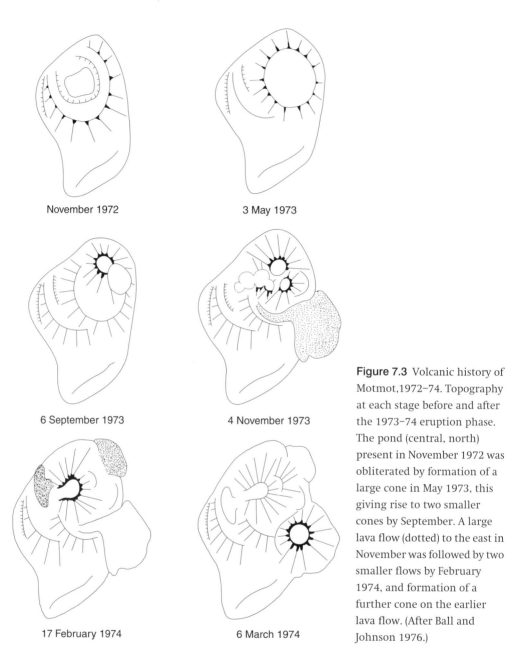

November 1972

3 May 1973

6 September 1973

4 November 1973

17 February 1974

6 March 1974

Figure 7.3 Volcanic history of Motmot,1972–74. Topography at each stage before and after the 1973–74 eruption phase. The pond (central, north) present in November 1972 was obliterated by formation of a large cone in May 1973, this giving rise to two smaller cones by September. A large lava flow (dotted) to the east in November was followed by two smaller flows by February 1974, and formation of a further cone on the earlier lava flow. (After Ball and Johnson 1976.)

'middle unit' in the stratigraphy, an unsorted ignimbrite layer. This unit is between 3 and 5 m thick at the coast, and includes carbonized and uncarbonized fallen tree trunks and branches. It overlies the western flank of Mt Réaumur as well as the eastern side of Long Island to a depth of over 30 m; it may have increased the area of the island, particularly to the south-east and north-west. Rocks of this extensive pyroclastic mantle are low-silica andesitic, whereas those of Réaumur and Cerisy volcanoes are basalts (Johnson *et al.* 1972). The final phase of each eruptive event was one of waning volcanic activity, and is represented on the ground by a shallower 'upper unit' of air-fall deposits (from about 1 to 2.5 m thick where it has not been lost by erosion), containing finer-grained (up to 1.5 cm in diameter), well-bedded, accretionary lapilli. Except in the case of the third event, this upper unit contains buried soils. All three eruptive sequences are thought to have been related to stages of caldera subsidence about 16 000, 2040 and 350 years ago, the last of which had far-reaching effects.

At the time of the seventeenth-century eruption the wind was from the north-east and the resulting high ash cloud was carried over to mainland New Guinea, where houses collapsed under the weight of ash (which also later produced new soils). An extensive characteristic ash-fall layer, known as the Tibito Tephra, covers an area of at least 87 000 km^2 over the Finisterre Range, Ramu–Markham, Jimi and Wahgi valleys, the Sepik–Wahgi divide and the Central Highlands, extending as far west as Tari and as far south as Lake Kutubu, Mt Karimui and Crater Mountain (Blong 1979, Pain and Blong 1979).

'The Time of Darkness' was originally (and erroneously) thought to have been associated with the 1883 eruption of Krakatau volcano, 4350 km to the west (Blong 1975). On the basis of radiocarbon dates of both the Tibito Tephra and the Long Island deposits, as well as evidence from analyses of the distribution of sulphate concentrations in cores of the Greenland ice field (indicating times of unusual sulphuric acid rain) the eruption is now dated as between AD 1640 and 1680, most probably around 1645 (Polach 1981, Zielinski *et al.* 1994), several decades before Dampier sailed past the 'long' island.

The Ritter Island event

About two centuries after the Long Island eruption, another event occurred about 100 km to the east, on another island of the Bismarck volcanic arc, now called Ritter (or Kulkul) (Fig. 7.4). The island lies about 10 km from the north-eastern coast of Umboi and 9 km south of Sakar. It is but the small remnant of a larger island volcano, said to have been almost 800 m high with a slope of 40° at its base. The island disintegrated catastrophically in 1888 (Blong 1982), losing over 95% of its bulk. The present island comprises only the eastern and south-eastern fragments of the former island's crater wall or caldera escarpment, and is uninhabited.

The unusually steep-sided volcanic cone that was Ritter's parent island had long been known as an active volcano under the name Volcano Island. In March 1700, whilst in command of HMS *Roebuck*, Dampier, one of the finest sailors of his day and an acute observer and accurate recorder, wrote that he 'saw a great Fire . . . blazing up in a Pillar, sometimes very high for 3 or 4 Minutes, then

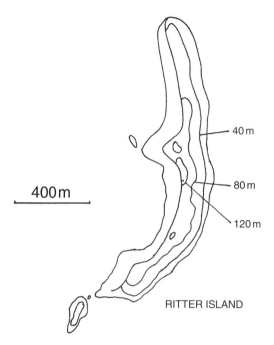

40 m

400 m

80 m

120 m

RITTER ISLAND

Figure 7.4 Ritter Island.

falling quite down for an equal space of Time; sometimes hardly visible, till it blazed up again.' Dampier's remarkably detailed description continued:

The Island all Night vomited Fire and Smoak very amazingly; and at every Belch we heard a dreadful Noise like Thunder, and saw a Flame of Fire after it, the most terrifying that I ever saw. The Intervals between its Belches, were about half a Minute: some more, others less: Neither were these Pulses or Eruptions alike; for some were but faint Convulsions, in Comparison of the more vigorous; yet even the weakest vented a great deal of Fire; but the largest made a roaring Noise, and sent up a great stream of Fire runing down to the Foot of the Island, even to the Shore. From the Furrows made by this descending Fire, we could in the Day Time see great Smoaks arise, which probably were made by the sulphureous Matter thrown out of the Funnel at the Top, which tumbling down to the Bottom, and there lying in a Heap, burn'd till either consumed or extinguished; and as long as it burn'd and kept its heat, so long the Smoak ascended from it; which we perceived to increase or decrease, according to the Quantity of Matter discharged from the Funnel.

Dampier's account suggests powerful Strombolian activity and perhaps pyroclastic flows, and other observers on ships passing through the strait (later named Dampier Strait) recorded eruptions in 1793 (with possible pyroclastic flows), and 1834, 1848 and 1878 (moderate vapour emissions) (Cooke 1981). There were also relatively weak vapour emissions in 1886 and 1887, and in early March 1888, less than 10 days before the catastrophe. A native of Sakar later said that before the event Ritter was uninhabited except for fishing parties, and was covered with vegetation.

On 13 March 1888 numerous earthquakes, which were felt on Sakar and New Britain, were followed by explosive activity beginning around 4 a.m., and shortly after dawn most of Volcano Island was disintegrated by a violent explosion (Taylor 1953). Large tsunamis wiped out coastal villages within a radius of 110 km, causing many casualties. On New Britain, some 20 km away, the wave was 12 m high and swept 800 m inland. There were many casualties on adjacent islands of course, particularly on Umboi and Sakar, and all animal life was expunged from several coral islets in the vicinity. Almost certainly all life on Volcano Island was destroyed.

On the assumption that the missing bulk of Volcano Island had been projected into the atmosphere, Lamb (1970) assigned to the event a 'dust veil index' (a measure of the opacity caused in the atmosphere) of 250. Cooke (1981), however, asserted that there is no direct evidence for a major explosion and suggested that the Ritter event probably contributed little in the way of dust veil or ash haze. He thought that sounds heard at Finshaven on New Guinea 150 km to the south and the 'fine, hardly noticeable' ash fall there could be accounted for by collapse alone. He noted that no noise, visible explosion, eruption column, incandescence, ash fall or earthquakes were reported by villagers and expedition members who survived the tsunamis in western New Britain and, indeed, he seriously doubted that there was an eruption at all.

There is no doubt, however, that there was a massive collapse of the cone, without warning. Cooke thought that this may have occurred in a single episode lasting no more than half an hour. About 2 km^3 of the island disappeared, leaving only a small section of the eastern side representing about a fifth to a quarter of the previous perimeter. The remaining fragment, Ritter, is steep-sided and arcuate, has an area of about 37 ha, is about 1800 m long and 200–300 m wide with the highest point 137 m above sea level. There is a sea stack off the island's southern tip and a smaller one midway along the western coast. On the west Ritter is bounded by two cuspate cliff scarps, part of a caldera wall, and from the arcuate shape of the present island one may surmise that the intact crater or caldera would have had a diameter of some 3 km. A Russian bathymetric survey around the island in 1974 confirmed the existence, west of the remnant, of a submerged collapse caldera about 2.5 km in diameter with a depth of more than 300 m below sea level (Cooke 1981).

After a period of 79 years of apparent inactivity (activity under water may not have been detected), there were brief explosive eruptions in the sea some hundreds of metres west of Ritter in 1972 and 1974 (Cooke et al. 1976). These eruptions were accompanied by swarms of strong and unusual earthquakes and small tsunamis which may have resulted from additional minor collapses of the caldera or from volcanic explosions (Cooke 1981).

It is almost certain that all life on Ritter was destroyed in 1888. Much of the present island is very steep, porous and barren but the eastern slopes are now covered in scrub and grass.

Biological recolonization of Long Island

The layer of hot ash deposited on Long Island's own forested surface was of Krakatau thickness –30 m or more in places – resulting in completely carbonized trees and probably destroying any freshwater aquatic life and most or all terrestrial life on the island. Emplacement temperatures of deposits of pyroclastic flows during the 1984 eruption of Mount St Helens were over 500 °C and those in a flow of St Augustine volcano in Alaska were 500–600 °C at depths of 3–5 m a few weeks after emplacement. In deposits that are several metres thick such temperatures may persist for several years. Seven years after they were deposited, poorly welded ash deposits in Alaska's Valley of Ten Thousand Smokes had fumarol temperatures as high as 645 °C (Fisher and Schminke 1984). It is very likely then, that Long Island's 1645 eruption destroyed most or all terrestrial life on the island. Long Island was recolonized by animals and plants from overseas during the following three and a half centuries.

The two islands nearest to Long Island are young, well-vegetated, extinct volcanoes: Tolokiwa, 41 km east of Long, is 46 km^2 in area and 1400 m high, and Crown, 10 km north-west of Long, is 14 km^2 in area and 570 m high. Umboi, about 80 km east-north-east of Long, is the largest (815 km^2) and highest (1650 m) island of the Vitiaz and Dampier Straits. It possesses a number of habitats; between an older mountain range in the east and a group of three forested younger cones in the centre is a shallow lake with extensive fringing marshland (Diamond 1974a). The two main sources of potential colonists to Long Island, however, are the large islands of New Guinea and New Britain, about 50 and 125 km distant, respectively. From his analysis of the bird fauna Diamond (1974a) concluded that the main source of colonists, somewhat surprisingly, was New Britain rather than New Guinea. Birds colonized New Britain from New Guinea, so that New Britain's avifauna has been already selected for good colonizing ability. Many of the species involved were categorized as 'tramps' or 'supertramps' (p. 115).

Currents in the Vitiaz Strait, between Long Island and New Guinea, flow south-eastwards for part of the year (Gorschkov 1974), at times of flood bringing surface flotsam to Long Island from the large estuaries of the Ramu and Sepik rivers on New Guinea's coast to the west-north-west. Franz Möder told Ball and Glucksman that rafts of floating vegetation were frequently washed up on Crown Island and, less often, on Long Island (Ball and Glucksman 1978: 466).

Long Island was resettled by people about 150–200 years ago (Ball and Hughes 1982, Specht *et al.* 1982) and its present inhabitants (and those of Tolokiwa) belong to a linguistic group that includes the people of Umboi Island and those of the adjacent Rai coast of New Guinea. Probably rather less than a thousand people now live in five small coastal villages, and have cleared large areas of lowland for gardens and coconut plantations. Feral pig and cuscus (*Phalanger* spp.) (introduced from Tolokiwa as a food source) are hunted with dogs in the forests of the caldera plateau, where both are common, but Lake Wisdom, which has no fish (p. 139), is not normally visited by Long Islanders.

Climate and vegetation

From records maintained on Umboi and on the New Guinea coast, as well as from comments made by district officers and administrators, Ball and Glucksman (1978) estimated Long Island's mean annual rainfall to be about 2800 mm. Rainfall is higher on the mountains than in the lowlands, and seasonal variation is greater than on the mainland; the wet season rainfall being less and the dry season longer. In the dry season (from April to November), winds can be strong and are predominantly south-easterly, but from about November to May they are from the north-west. The wettest months are from December to March, when thunderstorms are common and winds variable with north and north-west winds most frequent. The relatively low rainfall and 'over-drained' soils consisting largely of friable ash deposits result in fairly frequent droughts in prolonged dry seasons. At these times foliage may wither and turn brown, or leaf fall may occur (Ball and Hughes 1982). There was said to have been no rain at all for 8 months in 1972.

  We know that Long Island and its two smaller neighbours Crown and Tolokiwa were already forested when Dampier sailed through the straits in 1700, but nothing is known about the present floras of Crown and Tolokiwa, and until recently very little was known about that of Long Island, Previous biologist visitors (Coultas 1935, Evans 1940, Diamond 1970, Ball and Hughes 1982, Diamond *et al.* 1989) all concluded that Long Island's trees are fast-growing softwood species, dispersed either by floating propagules, such as those of *Barringtonia speciosa* and *Terminalia catappa* (both of which occur on Krakatau) or by birds or bats, like the seeds of *Ficus*, *Eugenia* or *Canarium* (Diamond *et al.* 1989). More than three decades ago Diamond (1970) believed that the lowland vegetation had not yet returned to its former composition and described it as an arrested subclimax. This description is still apt; the 300–400-m-high forest on the caldera plateau is somewhat open, with most trees appearing to be of the same age, clearly much younger than a few very large magnificent trees, such as strangler figs like *Ficus virens*, which may be centuries old and date from the time immediately following the eruption.

Birds

Species number

Diamond (1972, 1973, 1974a, 1974b, 1975, 1976, 1977) studied both the Long and Ritter cases as far as bird species are concerned, in 1972, and compared their

avifaunas and those of coral islets defaunated by the tsunamis with the avifaunas of unaffected islands in the area. The equilibria in numbers of bird species on both Long and Ritter, however, were lower than in the unaffected but otherwise similar islands, although the latter were low, and so might have been expected to carry smaller faunas. He found that on the devastated coral islets bird species numbers quickly returned to equilibrium, and by the 1970s the avifaunas on Ritter and the lowlands of Long had also reached an arrested equilibrium.

Diamond attributed the relatively species-poor new avifaunas on Long and Ritter to the fact that their lowland forest habitats were still at an immature stage of development. However, he noted that although Long's montane forest habitat had recovered well, he found fewer bird species there than in similar montane habitats on other high islands. Diamond concluded that montane birds generally had poorer dispersal powers than lowland species, and that their recolonization of the renewed habitat was therefore slower.

Recovery may have been constrained by the porous substrate and lack of streams, as Diamond suggested, as well as by intermittent ash fall (for not all subsequent eruptions may have been recorded). At high elevations, in the clouds above about 750 m on the two mountains where there are very strong winds, the montane forest is already structurally mature and Diamond believed that it had recovered more quickly than the lowland forest. Eldon Ball, however, suggested to me that it is more likely that both lowland and montane forest have reached an equilibrium climax stage, and the relative floral poverty of Long Island is due to the fact that, apart from the cloud-shrouded peaks, it is relatively dry.

Diamond *et al.* (1989) have noted that the avifauna has a number of characteristics of a 'new' fauna. There is an over-representation of 'supertramp' species (good-dispersing small-island specialists, absent from larger or older islands with large biotas – see below) and, correspondingly, a relative deficiency in specialist large-island 'high-S' species (see p. 116) that would be expected on an old island of Long's area. Moreover, whereas all Bismarck islands with an area of over $100\,km^2$ have recognizably distinct endemic subspecies, Long Island, with a land area of about $330\,km^2$, has none. In addition, Long Island has a quasi-equilibrium number of species, established before 1933, of about 50, which is well below the number expected for an island in this region of Long's size and isolation. The avifauna's characteristics are precisely those to be expected if the fauna was eliminated at the time of Long Island's eruption and has been reassembled since then. Diamond thought that the immaturity of the recovering lowland forest could be one explanation for the relatively low number of bird species. He believed that Long's avifauna has been selected for dispersal ability and that more slowly dispersing, high-S species would gradually arrive and, in time, displace the supertramps. However, although three more bird species have arrived since Diamond's 1972 survey, there is no evidence that supertramps are yet being displaced.

Table 7.1 *Numbers of species of selected plant and vertebrate groups recorded from Long Island and Motmot in 1999*

Taxon	Long Island	Motmot
Pteridophytes	17	7
Spermatophytes	145	38
Frogs	2	0
Crocodile	1	0
Snake	1	0
Varanid lizard	1	0
Geckos	5	0
Skinks	7	0
Resident land birds	50	3[a]
Pigeons and doves	8	0
Rats	2	0
Megabats	6	0
Microbats	2	1
Cuscus	1	0
Feral pig, dog, cat, chicken	4	0

[a] Further seven species recorded with no evidence of breeding, two only as prey remains.
Source: Thornton *et al.* (2001).

Analysis of the avifauna led to the conclusion that most immigrants were probably pre-adapted colonists from New Britain (which they had colonized from New Guinea), rather than from the larger and closer source, the island of New Guinea. Diamond *et al.* (1989) regarded Crown, Long and Tolokiwa as the 'Long group'; all three have avifaunas that differ in several respects from what one would expect on these islands had they not been defaunated by Long's eruption. Crown's bird fauna is a subset of that of Long; all its 33 non-migratory land bird species also occur on Long, as do 36 of Tolokiwa's 44 species.

Prior to 1999, only the birds of Long Island's terrestrial biota had been surveyed (Coultas 1935, Diamond 1972, 1976, 1977, 1981, Diamond *et al.* 1989), and two important avifauna datum points have been established, at 1933 and 1972, with which future surveys can be compared.

Our team of biologists surveyed the western parts of Long and the emergent island, Motmot (see below), for plants (especially figs), birds, reptiles, mammals and invertebrates, including fig wasps, over 15 days in June and July 1999. We found 145 species of seed plants (including 31 *Ficus* species: Shanahan *et al.* 2001) and 17 pteridophytes, to total 162 vascular plants (Harrison *et al.* 2001) on Long Island (Table 7.1).

Table 7.2 *Number (with percentage in parentheses) of resident land birds in various feeding guilds present on Anak Krakatau (1996) and Long Island (1999, taking the fauna as 50 species)*

Feeding guild	Krakatau	Long
Aquatic	0 (0)	2 (4)
Predator vertebrates	6 (17)	7 (14)
Predator invertebrates, small vertebrates	1 (3)	5 (10)
Specialist frugivore	4 (11)	4 (8)
Frugivore–granivore	1 (3)	2 (4)
Frugivore–insectivore	6 (17)	3 (6)
Aerial insectivore	6 (17)	5 (10)
Other insectivore	7 (19)	8 (16)
Ground insectivore–herbivore	2 (6)	6 (12)
Granivore	0	2 (4)
Omnivore	0	1 (2)
Nectarivore	3 (8)	5 (11)
Total species	36	50

Source: Schipper *et al.* (2001).

We assess the present non-migrant land bird fauna of Long Island as about 49 species (Schipper *et al.* 2001) (Table 7.2). One of the species recorded in Coultas's survey in 1933, the eastern black-capped lory, *Lorius hypoinochrous*, appears to have become extinct. Three bird species have colonized and four have become extinct in the past 27 years, a turnover rate of 0.26% of the fauna per year, about the same as the rate for Krakatau (Thornton *et al.* 1990a, 1993). The quasi-equilibrium noted by Diamond in 1972 as having lasted since 1933, now counted as 49–50 species (13–14 below the predicted equilibrium number of 64), is persisting, and I believe may continue for decades.

It appears that there has not yet been time for the island's forest to mature sufficiently to allow any of the 11 supertramps to be replaced by later-colonizing species. The proportion of supertramps has not changed in 27 years and the island's species number remains at a level (about 50 species) well below the expected equilibrium number of 64. Diamond suggested that this low number may be explained by Long Island's highly porous substrate and consequent immature flora. We believe that these conditions may persist for many more decades, with the species number tracking, not the theoretical equilibrium number of 64, but a lower, practical equilibrium achievable in the conditions prevailing on Long Island. Long Island may continue to be 'unsaturated' in theory but 'saturated' in practice for a long period.

Guild spectrum

The small number of aquatic bird species (two) on an island with a large lake may be explained by the lake's youth and isolation, but also by the fact that its shores, being caldera walls, are mostly steeply falling below water, providing few shallow marginal areas and thus a paucity of vascular waterplants (Ball and Glucksman 1978). Long Island also has only two granivores; compared to forest, grassland source areas on New Britain and New Guinea are more distant and relatively small and species-poor. In contrast, New Guinea is rich in ground-feeding forest birds, obligate frugivores and nectar-feeders. For example Beehler *et al.* (1986) enumerated 25 species of *Ptilinopus* and *Ducula*, specialist soft-fruit pigeons, and 21 lory species, whose fringed tongues are specially adapted for taking nectar. These guilds are well represented on Long Island.

It is of interest to compare the guild spectrum of Long Island's avifauna with that of the 117-year-old avifauna of Krakatau, which is probably only a third of the age (Table 7.2). If one excludes Long Island's two aquatic species from the comparison (Krakatau has no permanent freshwater body), notable differences are that Long Island has greater proportions of predators of invertebrates and small vertebrates, specialist frugivores, insectivore–granivores and nectarivores, and a smaller proportion of the important generalist frugivore–insectivore guild. So although the replacement of supertramps is much slower on Long Island than is believed to be the case more generally, Long Island nevertheless has a more specialized feeding guild spectrum than does Krakatau. The special conditions on Long appear to have delayed the replacement of supertramps by high-S species but have not affected the change from generalist to specialist feeders. It is possible that these conditions are very long-lasting, with the island's species number tracking, not the theoretical equilibrium number for a Bismarck island of Long's size, but a practical equilibrium achievable on the island under the present conditions. Thus the equilibrium on Long Island should perhaps be regarded as that appropriate to the island, Long being yet another exception to the strict application of island equilibrium theory.

Character divergence on Long Island

Jared Diamond (Diamond 1976, Diamond *et al.* 1989) drew attention to an interesting case of an evolutionary process known as character divergence on Long Island. It involves two myzomelid honeyeaters, and has become a classic example of the rapid evolution of character divergence. Two species of *Myzomela* coexist on Long Island: the smaller *Myzomela sclateri*, of the southern Bismarcks, and, surprisingly, the larger species, *Myzomela pammelaena*, which is otherwise confined to the northern Bismarcks, some 350 km to the north. The two species occur together only on the 'Long group' (Long, Crown and Tolokiwa), and of nine congeneric bird species pairs occurring on Long Island, this is the only pair occuring together only on Long Island (and so probably first met there) and both species are morphologically different from their conspecifics elsewhere. On the Long group *M. pammelaena* is even larger than it is elsewhere, and *M. sclateri* is even smaller than it is in the rest of its range. The two species are thus more different in size where they co-occur than they are otherwise, and from what is known of other sympatric honeyeaters, Diamond's group concluded that this represents character divergence in response to the presence of the other species of the pair, permitting them both to persist in the same habitat on the Long group. *Myzomela pammelaena* on the Long group is well out of its normal geographical range, and it seems highly likely that both honeyeaters have colonized Long Island since the mid-seventeenth-century eruption. If so, then the evolution of character divergence has occurred in only three and a half centuries.

Apart from the birds, some 29 other land vertebrate species are now known from Long Island, two of them amphibians, and at least ten indigenous mammal species – two rats, eight bats, including six pteropodids (two flying foxes, a tube-nosed fruit bat and three blossom bats), and two insectivorous microbats (an emballonurid and a hipposiderid). The cuscus was probably introduced and pigs, dogs, cats and fowl are feral.

Amphibians

The two amphibian species known from the island are frogs: the hylid tree frog *Litoria infrafrenata* and the ranid *Platymantis papuensis. Litoria infrafrenata* is one of the largest tree frogs in the world, being up to 13.5 cm in length. Of the 80 or so hylid frogs in New Guinea, only three, including this one, reach New Britain. For tree frogs, therefore, this species can be said to be a good colonizer, and has one of the main attributes of a good colonist – wide ecological tolerance (see below).

Platymantis papuensis is very common, and the islanders believe it was intro-duced (Diamond 1989). It is also 'one of the most abundant frogs in the lowlands [of New Guinea] north of the central dividing ranges' (Menzies 1975: 26). Colour varieties on Long Island include pink, grey and mottled, and in New Guinea the commonest is plain brown, another has a light mid-dorsal band, and yet another has two dorsolateral light bands. This is a terrestrial frog; the toes are not at all webbed; the large, few eggs are laid on land in concealed damp places; and the free tadpole stage is absent, metamorphosis taking place within the egg, from which young frogs emerge directly (Menzies 1975). This frog is therefore also well adapted as a potential colonist.

Reptiles

At least ten species of land reptiles and three marine turtles occur on Long Island.

The land reptiles are seven to nine species of skinks: two or three species each of *Sphenomorphus* (two of which also occur on Karkar, one of them also on Umboi) and *Emoia* (one of which occurs also on Karkar and Crown), two of *Eugongylus* (both of which occur on Karkar, one also occurring on Crown and Umboi, the other also on Tolikawa), and *Basia smaragdina* (present also on Crown). The island's single snake species, the Pacific tree boa, *Candoia carinata*, is very common, and the mangrove monitor, *Varanus indicus*, is hunted for its skin, which is used for drum-heads. Other reptiles include the estuarine crocodile, *Crocodylus porosus*, which is believed to be still present in Lake Wisdom (although last seen over a decade ago; see Anderson 1978/9), and four or five gecko species: the house gecko, *Hemidactylus frenatus*, *Cyrtodactylus* cf. *pelagicus*, *Lepidodactylus* cf. *lugubris*, and possibly another species of *Lepidodactylus*, and a species of *Gehyra*. Three species of marine turtles are known to use the island's beaches: the green turtle, *Chelonia mydas*, the hawk-billed turtle, *Eretmochelys imbricata* and the leathery turtle, *Dermochelys coriacea*.

Mammals

The eight known species of bats include the variable flying fox, *Pteropus hypomelanus* (found also on on the islands of Karkar, Bagabag and Umboi), the giant flying fox, *Pteropus neohibernicus neohibernicus* (Karkar, Umboi, Sakar) and at least four other pteropodids, the Bismarck tube-nosed bat, *Nyctimene vizcaccia* (Umboi and possibly Karkar, Bagabag, Crown, Tolokiwa, Sakar) and three macroglossine nectar or blossom bats: the least blossom bat, *Macroglossus minimus nanua* (Karkar, Crown, Tolokiwa, Umboi, Sakar), black-bellied bat, *Melonycteris melanops* (Tolokiwa, Umboi) and common blossom bat, *Syconycteris australis papuensis* (Bagabag, Crown, Tolokiwa, Umboi, Sakar). Two microbats have also been recorded on Long Island, the lesser sheath-tailed bat, *Mosia nigrescens solomonis* (Crown, Umboi) and the fawn leaf-nosed bat, *Hipposideros cervinus cervinus* (Karkar, Crown) (Koopman 1979, Bonaccorso 1998, Cook *et al.* 2001). All eight species also occur on New Britain and New Ireland, all but one (*M. melanops*) are found on Manus (Admiralty Islands), and all but two (*N. vizcaccia* and *M. melanops*) occur on New Guinea.

The food of *Pteropus hypomelanus* includes figs, *Terminalia* fruit, kapok and coconut flowers as well as cultivated fruits like banana, pawpaw and mango. The species can commute over long distances to feed.

Pteropus neohibernicus, weighing about 1.6 kg as an adult, is one of the world's largest bats. Bonaccorso (1998) noted that fruits of *Ficus*, *Terminalia* and *Calophyllum* are favourite foods, and blossoms are also eaten throughout the day, although activity increases after dark.

Nyctimene vizcaccia is a fig-eater that is found on small- to medium-sized islands in closed canopy forest, rarely venturing into secondary growth or plantations. This bat (with flying foxes and fruit-eating birds) could have been instrumental in the colonization by fig species of Long Island from Umboi, and of Motmot from Long Island (see below).

Syconycteris australis occurs in a wide range of forest and woodland habitats. In New Guinea fruits form most of its diet and small seeds of several families, including figs (Moraceae), Piperaceae and Solanaceae, have been found in its faeces. The species is not active until well after sunset and has been recorded as occasional prey of the sooty owl (*Tyto tenebricosa*) in New Guinea, so it may be taken by the barn owl (*Tyto alba*) on Long Island. This fourth bat could well be a disperser of *Ficus* species, at least from Umboi to Long, and within Long.

The upper surface of the long, slender, highly protrusible tongue of *Macroglossus minimus* is covered with rows of backwardly directed scale-like papillae and the tip with bristle-like papillae. It is not a primary forest inhabitant. Its food is almost entirely nectar and pollen together with some insects, and no fruit seeds have been found in the stomach. This bat is a pollinator rather than a disperser of plants.

Melonycteris melanops is New Guinea's largest blossom bat and is an important pollinator of wild bananas. Like the previous species it is common in gardens, plantations and disturbed habitats, rare in primary forest (Smith and Hood 1981, Bonaccorso 1998), and like *P. hypomelanus* it has a low metabolic rate compared to other mammals of its size (Bonaccorso and McNab 1997).

In contrast, weighing as little as 2.5 g, *Mosia nigrescens* is one of the smallest of bats. It begins to feed in the late afternoon (Flannery 1995) on aerial insects and those clinging to foliage from the canopy to near ground level, roosting in caves and under overhangs, including large-leaved plants. Bonaccorso (1998) found four individuals under a fish-tail palm frond, stacked on top of one another belly-to-back, possibly reducing heat loss. Roosting bats are alert throughout the day.

Hipposideros cervinus feeds on aerial insects and also gleans in dense vegetation, roosting in caves, mines, tunnels, deserted houses and tree hollows. It has a basal metabolic rate of only 60% of what would be expected from its body mass. The species is taken in flight by *Falco severus* (Oriental hobby), and constricting snakes such as pythons may be important predators in the day roosts (Bonaccorso 1998).

Thus, five of the eight species of bats on Long Island (the two flying foxes, the Bismarck tube-nosed bat and two macroglossines, the black-bellied bat and common blossom bat) are possible dispersers of figs (see below). One of the flying foxes, *P. hypomelanus*, with a low energy budget (perhaps an adaptation for living on low-resource islands) is a supertramp (in the sense of Diamond), and two other pteropodids, *N. vizcaccia* and *M. melanops*, may be designated at least

Table 7.3 *Summary of ecological features of Long Island bats*

	Fig-eating	Wide-ranging	Eurytopic	Supertramp
Pteropus hypomelanus	●	●	●	●
Pteropus neohibernicus	●	–	●	–
Nyctimene vizcaccia	●	●	–	●
Macroglossus minimus	–	●	●	–
Melonycteris melanops	●	–	●	●
Syconycteris australis	●	●	●	–
Mosia nigrescens	–	●	●	–
Hipposideros cervinus	–	●	●	–

Source: After Bonaccorso (1998).

tramps, if not supertramps (see below). Three other bats, *S. australis*, *M. nigrescens*, and *H. cervinus*, are habitat generalists (Table 7.3).

The Polynesian rat (*Rattus exulans*), which has been carried in canoes to many Pacific islands, is present (it has also been recorded from Tolokiwa), along with the spiny rat, *Rattus praetor* (also on Karkar). The oriental cuscus, *Phalanger orientalis*, a loris-like arboreal marsupial, is abundant, and also known from Karkar, Bagabag, Tolokiwa and Umboi. We were told by the local people that the first breeding pair was introduced to the island by a couple from the Siassi Islands in the late 1940s, although other reports from Long Islanders state that the animal was introduced from Tolokiwa (Diamond 1974a, Koopman 1979). It is now hunted regularly, an average of five to six animals being taken at each hunt. Meat is consumed locally and some animals are sent to Madang market for fur and meat. Feral pigs, *Sus scrofa*, are also common and hunted with dogs. Feral chickens, *Gallus gallus*, are not taken for food – there is no shortage of protein on Long Island. The mammal fauna (11 species) thus has all the characteristics of one that has been assembled over a considerable sea barrier – largely ecologically tolerant or 'tramp' bats, the only ground-living mammals being those that have probably been carried to the island, intentionally or not, by people.

Ficus species

One might think that of all tree species figs would be poor colonists, for each must have its own particular wasp species present to effect pollination and thus establishment. Yet there are now at least 31 fig species on Long Island. In comparison, on the Krakataus, after over a hundred years of recolonization, there are some 22 fig species.

Sections of the genus are represented on Long Island in much the same proportions as in Madang Province, except section *Rhizocladus*. This group of

Table 7.4 *Comparison of proportions of* Ficus *species with monoecious (M) or dioecious (D) breeding systems on Motmot, Long Island and the mainland of New Guinea, and on Anak Krakatau and the Krakatau archipelago in the first 25 years after 1883*

Locality	Breeding system	Number of species	Percentage of all *Ficus* species
New Guinea	D	109	74
	M	38	26
Madang Province	D	38	69
	M	17	31
Long Island	D	21	67
	M	10	33
Motmot	D	6	75
	M	2	25
Krakatau by 1918	D	7	100
	M	0	0
Anak Krakatau	D	6	86
	M	1	14

climbers makes up 20% of New Guinea's 147 *Ficus* species but no representatives were found on Long Island. The relative youth of Long Island's forests, with a dearth of large mature trees as substrates, is one possible explanation. Another, and more likely, is the west–east decline in species of this section of the genus across New Guinea, only four being known among the 55 *Ficus* species of Madang Province (7%), for example. Long Island has the same proportions of monoecious and dioecious fig species as Madang Province, which is the same as New Guinea as a whole (about 30% monoecious) (Table 7.4); there is no evidence from this that either breeding system increases the chances of successful colonization.

Over 30 species of potentially or known fig-eating vertebrates occur on Long Island, and there are records of 21 of Long Island's fig species being eaten, either on the island or elsewhere.

Until recently it was thought that the passage time of small seeds in the alimentary tract of bats (they spit out large seeds) was too short for them to be carried over distances of 50 km within the bat. Shilton *et al.* (1999), however, reported that when confined in small cages small *Cynopterus* fruit bats retain seeds in the gut for up to 12 hr; if this is shown to be possible in naturally exercising fruit bats, then bats, as well as frugivorous birds, may play a more important role in the dissemination of fig seeds, for example, than was thought. Both of Long Island's flying fox species could have carried fig seeds to Long Island from sources on Umboi or perhaps from New Guinea.

The two species of large flying foxes (*Pteropus* species) found on Long Island may well have been responsible for bringing viable fig seeds to the island in their guts (see above). Figs make up a large part of the diet of both species. The supertramp bird species of Long Island include the yellow-bibbed or Solomons fruit dove and the grey imperial pigeon, both of which are very common and excellent dispersers of fruiting trees, including figs, and there are three other species of these genera among Long Island's ten species of pigeons and doves, some of which are also extremely common. There are two species each of *Ptilinopus* fruit doves, *Ducula* imperial pigeons and *Aplonis* starlings, and these were perhaps more likely agents of fig dispersal to Long Island than fruit bats. Almost certainly the four species of larger imperial pigeons and fruit doves, which are strong flyers that retain viable seeds for long periods, were involved. Species of all these genera were the likely dispersers of fig species to Krakatau after its 1883 eruption (Thornton 1996a).

Sixteen *Ficus* species on Long Island were found to have ripe figs during our 1999 visit. All these had their pollinating wasps, and it is likely that all Long Island's figs have their pollinators and are capable of dispersing and becoming established within the island, for a range of smaller dispersers is present on Long Island to assist this. Of these, six species are probably the most important, three of them bats: *Nyctimene vizcaccia*, which is 'a dietary specialist, feeding primarily on figs' (Bonaccorso 1998: 182), *Syconycteris australis*, which in New Guinea feeds mainly on fruit, fig seeds having been found in its faeces; and *Melonycteris melanops*, which takes nectar and pollen but has eaten soft fruit in captivity (Bonaccorso 1998). The Oriental cuscus, Louisiades white-eye, *Zosterops griseotincta*, and common koel, *Eudynamys scolopacea*, are also fig-eaters, making 14 potential dispersers in all. *Macropygia* cuckoo-doves and rainbow lorikeets, *Trichoglossus haematodus*, are considered to be seed predators rather than dispersers.

The nature of the colonists
Supertramps
Diamond (1974a, 1975) had come to know the bird species of the western Pacific well, and was able to assign to them dispersal-strategy categories. He allocated to each species a 'colonization index', which is an assessment of its chances of colonizing islands of certain types and with certain faunal spectra. He also devised a set of 'assembly rules', largely based on competitive exclusion networks and hierarchies, rules that are applicable to the build-up of island avifaunas in this region (and to a much less extent to continental Australia for example: Diamond 1975). Diamond designated as 'supertramps' those species that specialize in rapid breeding and over-water colonization, are found only on

small or unstable species-poor islands where resource levels are high, and are excluded from larger or older islands with a high number of bird species. Their complement, 'high-S species', are those which flourish in competitive, low-resource situations involving a large number of species. Long Island has an over-representation of the former and an under-representation of the latter.

Of some 14 supertramp bird species identified in the Bismarcks, 11 (a fifth of its entire bird fauna) occur on Long Island. Ten of them – Mackinlay's cuckoo-dove, *Macropygia mackinlayi karkari*; Nicobar pigeon, *Caloenas nicobarica*; yellow-bibbed fruit-dove, *Ptilinopus solomonensis speciosus*; grey imperial pigeon, *Ducula pistrinaria*; collared kingfisher, *Halcyon chloris*; island monarch, *Monarcha cineras-cens*; mangrove golden whistler, *Pachycephala melanura*; Bismarck black honey-eater, *Myzomela pammelaena*; scarlet-bibbed honeyeater, *Myzomela sclateri*; and Louisiades white-eye, *Zosterops griseotincta* – were already established by 1933 when Coultas visited the island. Another, the pied imperial pigeon, *Ducula bicolor*, had arrived by 1972, replacing a species which was not a supertramp, the eastern black-capped lory (Diamond *et al.* 1989). Nine of the supertramps were widespread on the island in 1972, occupying all habitats, and densities were from four to 11 times those on more distant control islands (Diamond 1970, 1974a, 1974b). So, by 1972 Long Island's 11 bird supertramps made up about 22% of its avifauna, a very high proportion and more than three times the number expected for islands with as many bird species as Long Island. Umboi, for example, has only four supertramps. In contrast, Long Island has only two high-S species, whereas Umboi has 16.

Supertramps may also be recognized among the bats. According to Bonaccorso (1998), the variable flying fox (*P. hypomelanus*) is a 'supertramp' *sensu* Diamond (1975) in New Guinea, occurring on many species-poor small islands throughout its wide range and rarely on large islands and continents, which are avoided as roosting sites even when within visible range. It has been collected from New Guinea Island only twice, and occurs regularly on mainlands only in the south-east coast of Thailand, Cambodia and Vietnam. This bat has an unusually low basal metabolic rate and low energy budget compared to other bats and other frugivo-rous mammals of its size, and Bonaccorso suggested that this may be an adapta-tion permitting the species to survive on small, resource-poor islands. Two other Long Island bats may be designated as supertramps, at least in the Bismarcks region. These are the black-bellied bat, which is absent from the main island of New Guinea, and also has a relatively low basal metabolic rate (Bonaccorso 1998) and the Bismarck tube-nosed bat, which prefers small-to-medium-sized islands, and also is absent from New Guinea Island. Thus three of a known bat fauna of eight species are supertramps, an even higher proportion than in the birds.

Supertramps are important early colonists of islands but cannot survive the interspecific competition that comes with more species, and are expected to be replaced, as Diamond (1981) predicted for the birds, by later, high-S species immigrants. This has not yet occurred. We cannot make statements about the bats, which have only been surveyed recently, but in the seven decades since bird surveys began, no bird supertramp has become extinct, turnover has been very low, and the avifauna has still not achieved an equilibrium number appropriate to an island of Long Island's area in the region. Evidently there has not yet been time for the island's forest to mature sufficiently to allow any of the 11 bird supertramps to be replaced by later-colonizing species. The proportion of supertramps has not changed in 27 years and the island's species number remains at a level (about 50 species) well below the expected equilibrium number of 64. Diamond (1974a, 1974b, 1976) suggested that this low number may be explained by Long Island's highly porous substrate and consequent immature flora. I believe, along with Diamond (1974a, 1974b, 1976) and Ball and Hughes (1982), that the vegetation of Long Island is being held at a sub-climax stage by some physical factor, probably the poor water retention of the porous, ashy substrate, in turn holding the avifauna below the theoretical species number and allowing the high proportion of supertramps to persist.

Colonists with broad ecological tolerance

As was found for the animals of the Krakatau Islands (Thornton *et al.* 1990a, Thornton 1996a), where the biota is even younger, only a little more than 100 years old, a number of Long Island's successful colonists have a wide geographic distribution and broad ecological tolerance.

Of the four species of freshwater molluscs known from Long Island, the commonest in 1999 was the small, turret-shelled *Melanoides tuberculatus*, found in abundance on algae in the first survey of Motmot in 1969 (Bassot and Ball 1972) and occurring in Lake Wisdom in mud or algae at all depths down to about 350 m (Ball and Glucksman 1978). This small species may be dispersed by birds; it is taken by ducks (Ball and Glucksman 1975). Possibly originating in the West Indies, it is now found throughout the tropics. It is present in Lake Dakataua in New Britain and is a colonist of the two very small bodies of freshwater on the Krakataus. Bassot and Ball (1972) note that the species can live in temporary ponds, streams, lakes, rivers and in hot or brackish waters.

Of the two amphibian species known from Long Island, the hylid *Litoria infrafrenata* is found in the far north of Queensland, Australia, and is the most widely distributed of all the hylid frogs in the Papuan Region below about 1200 m altitude, ranging from eastern Indonesia to the Bismarcks, Admiralties and Louisiades (Menzies 1975). Menzies noted that although it is a forest species, preferring to breed in deep, shaded swamps, it is also among the few tree frog species that inhabit gardens and suburban land in New Guinea.

Platymantis papuensis is found below about 800 m altitude from eastern Indonesia through the north coast lowlands of New Guinea to Milne Bay, the d'Entrecasteaux and Trobriand Islands. Once again Menzies noted that 'although it is a forest frog, it can also be found in gardens where there is extensive shrubbery and long grass' (Menzies 1975: 26).

The megapode on Long Island (and Motmot), *Megapodius eremita*, is practically omnivorous and, as noted below (under Motmot), has the widest range of breeding habitats of any megapode.

Seven of the eight known Long Island bat species are eurytopic, and most of these are wide-ranging. The two flying foxes, common blossom bat, *Syconycteris australis*, lesser sheath-tailed bat, *Mosia nigrescens* and fawn leaf-nosed bat, *Hipposideros cervinus* all tolerate a wide range of habitats. The least blossom bat, *Macroglossus minimus* and *Melonycteris melanops* are respectively absent from or rare in primary forest (Bonaccorso 1998).

One of Long Island's flying foxes, *Pteropus hypomelanus*, ranges from Thailand to the Philippines, usually up to 500 m altitude (but in the Philippines to 900 m) and forages in 'primary and secondary lowland and hill forest, small gardens, and plantations' (Bonaccorso 1998: 127). The other, *P. neohibernicus*, is not known outside the New Guinea region, but occurs in a wide range of forest types from sea level to 1400 m altitude. *Syconycteris australis* is found from Ambon to north-eastern Queensland, and Bonaccorso (1998: 205) described it as 'a habitat and feeding generalist' which 'feeds on nectar, pollen, fruits, and perhaps small quantities of insects'. It occurs in primary and secondary montane, moss, hill and lowland rainforest, swamp forest and dry sclerophyll woodland. Bonaccorso (1998) described *M. nigrescens* as a 'habitat generalist, foraging in primary and secondary tropical broadleaf forests, mangroves, gardens, coconut plantations, and villages' (see also below, under Motmot). *Hipposideros cervinus* has a range extending from Malaysia and the Philippines to Vanuatu and Australia, and is widespread throughout the New Guinea region. It is a 'habitat generalist found in both primary and secondary rain forest, eucalyptus woodlands, gardens, plantations and urban areas up to lower montane elevations' (Bonaccorso 1998: 273). *Macroglossus minimus* is another species with a wide geographical range, extending from Thailand and the Philippines to the Solomons and Australia. It is common in lowland secondary rainforest, *Melaleuca* woodlands, mangroves, gardens, plantations and towns, but is uncommon in primary rainforest. *Melonycteris melanops*, endemic to Papua New Guinea, is common in gardens and plantations and is rare in primary forest (Smith and Hood 1981, Bonaccorso 1998). Only one of Long Island's eight bat species has a strong habitat preference for true closed-canopy forest – *Nyctimene vizcaccia*.

The two rat species present are the Polynesian rat, *Rattus exulans* and the spiny rat, *Rattus praetor*; *R. exulans* has been carried in canoes to many Pacific islands and is present on nearby Tolokiwa. It is the only rodent to benefit from slash-and-burn agricultural practices (Dwyer 1978, 1984). *Rattus praetor* occurs on Tolokiwa and Karkar.

A tighter focus

Where the devastation of an island has been more recent, it has sometimes been possible to follow the assembly process from a zero or near-zero base-line.

Thus the cases of Volcano, San Benedicto, Thera and Long Island all suffer from disadvantages in so far as studies of their ecology are concerned. The destructive events on Thera and Long Island, on which eradication of the biota was probably complete, occurred too long ago for biotic recovery to have been monitored by scientists. The events on Volcano and San Benedicto were recent, and their effects and the recovery (of plants on Volcano and plants and animals on San Benedicto) monitored, but in each case the destruction was incomplete; in San Benedicto's case about half the biota survived the eruption.

The great 1883 eruption of Krakatau provided a case which lacked both of these drawbacks. Extirpation of at least the macrobiota of Krakatau was thought to be complete, and recovery could be monitored.

CHAPTER EIGHT

Krakatau, Sunda Strait

In 1883 the 7-km-long, 800-m-high island of Krakatau, in Sunda Strait, Indonesia (Fig. 8.1), erupted explosively with the force of more than 10 000 Hiroshima-type atomic bombs. Two-thirds of the island, including the two volcanoes that were active and half of the third and highest volcano, Rakata, now being unsupported, slumped into the emptied magma chamber, forming a 200-m-deep submerged caldera.

The remaining third of Krakatau, comprising half of the Rakata volcano, now known as Rakata Island, and the two closely adjacent islands, Sertung and Panjang (Fig. 3.3, p. 38), were covered in a blanket of hot ash some 30 m thick, thicker in places, which it is believed extirpated the animals and plants, thus setting in train a long-running, large-scale, 'natural experiment' in biotic colonization. The Krakataus are some 44 and 35 km from the biologically rich islands of Java and Sumatra, respectively, and the over-sea dispersal and recolonization of these islands by animals and plants, the course of the primary succession that followed, and the assembly of a new tropical forest community have been studied over the ensuing 110 years, with variable regularity and intensity (see, for example, Docters van Leeuwen 1936, Dammerman 1948, Tagawa *et al.* 1985, Whittaker *et al.* 1989, Thornton *et al.* 1990b, Thornton 1992a, 1996a, 1996b).

Monitoring of the reassembly of a flora on the three islands began in 1886 (Treub 1888) and continued fairly regularly until the 1930s (Docters van Leeuwen 1936). Van Borssum Waalkes (1960) made an important survey of the plants of Rakata in 1951. In contrast to the early monitoring of floral assembly, however, the first faunal survey was not until 1908 (Jacobson 1909), 25 years after the eruption, but there were several zoological surveys between 1919 and 1934 (Dammerman 1948). During the next 45 years the only animal survey of the three islands was by Hoogerwerf (1953b) in 1951–52, and concerned only the birds. Since 1979 there have been a number of animal and plant surveys (see, for example, Yukawa *et al.* 1984, Tagawa *et al.* 1985, Ibkar-Kramadibrata *et al.* 1986, Thornton and New 1988b, Whittaker *et al.* 1989, Thornton *et al.* 1990b, Bush and Whittaker 1991, Thornton 1991, 1992a). Before reviewing the reassembly of a biota on the devastated islands, however, it is necessary to consider the question: what was the biota before the 1883 eruption?

Figure 8.1 The Sunda Strait, between Sumatra and Java, Indonesia, to indicate position of the Krakatau archipelago, the possible stepping stone island of Sebesi and other localities mentioned in the text.

The pre-1883 biota

Prior to the eruption, specialist collectors visited Krakatau rarely; only six or seven species of vascular plants and five fairly large and conspicuous land mollusc species were known. From the reports of occasional visitors such as geologists and from ships' logs, however, we at least know that before the eruption the islands were forested from their shores to the 800-m-high peak of Rakata Volcano. Although the forest botanist Richards (1952) suggested that the pre-1883 forests were probably similar to those of the Sumatran and Javan mainlands, there is geological evidence that the present islands may be the remnants of a larger, prehistoric island volcano, 'Ancient Krakatau', and if so their biotas would probably have been somewhat disharmonic, lacking at least some poorly dispersing mainland species, such as some of the large tree species of forest interiors.

Dammerman (1948) tried to assess the nature of Krakatau's pre-1883 biota by comparing its present biota with those of otherwise similar, unaffected islands of the region. The island of Sebesi is some 13 km to the north, is volcanic, and about the same height and size as the present island of Rakata. However,

Sebesi's own biota was substantially depleted as a result of the Krakatau eruption and for several decades its lowlands have been covered in coconut plantations, only the summit area above about 700 m surviving in a natural condition. It is therefore far from an ideal island for comparative purposes (see p. 218). Nevertheless, in 1921 Docters van Leeuwen collected 359 vascular plants on Sebesi compared to the 259 species found on the Krakataus at that time, and clearly it suffered less destruction and recovered more quickly than the Krakataus. It is also closer (about 14 km distant) to the rich potential source area of Sumatra. Dammerman (1948) also found that in 1920–22 Sebesi's fauna, although similar in size (638 animal species compared to 621 on Rakata), was the richer in snakes, moths, Orthoptera (twice as many), Thysanoptera (then not recorded from the Krakataus), cicindelid (tiger) beetles (six to zero), myriapods (including two families not present on the Krakataus), earthworms (five to two) and land molluscs (more than three times as many species, including three slugs and six snails not present on the Krakataus). He ascribed these differences to Sebesi's greater proximity to a mainland source and the more rapid rate of recovery of its vegetation, and concluded that the extent of faunal devastation in 1883 had been almost the same as it was on Rakata.

Dammerman therefore chose the island of Greater Durian as a more appropriate example of the Krakataus' original fauna (Thornton 1992b). Similar to Rakata in size and occurring in the same general region, Durian is isolated from large land masses to about the same extent (35 km from Sumatra and 60 km from Singapore) but there are more possible intervening stepping stone islands from potential sources. At 309 m, it is lower than Rakata, the soil is poor and the vegetation impoverished, including a near absence of fruit-bearing trees. Like Sebesi and unlike Krakatau it has fresh water and is inhabited.

Dammerman acknowledged that Durian, too, provides far from the ideal comparison. In 1923 Durian's diversity was higher than Rakata's in mammals (particularly bats), resident land birds and reptiles (particularly snakes) and there was a small freshwater vertebrate fauna. There were more insect species (738 compared to about 500 on Rakata at that time), and a much greater diversity of moths but about the same number of butterfly species. Also richer on Durian were orthopteroid insects, which had a much greater representation of acridids, crickets and mantids; termites, which included mound-building species; and Neuroptera. Apart from in moths and Orthoptera, and in several non-flying or freshwater vertebrate groups which would not be expected to have the same diversity on the two islands, the present Krakataus' fauna appears to have reached, and in many groups exceeded, the diversity of Durian, further evidence of the latter's impoverishment.

Thus neither the comparison with Sebesi nor that with Durian were very useful, for different reasons. The best estimate of the *minimum* size of Rakata's pre-1883 biota (those of Sertung and Panjang assumed to be subsets of this) may be its present biota. The number of vascular plant species now present is about 360 (over 400 on the archipelago).

The first colonists

What were the first colonizers of the devastated Krakataus after the 1883 event, and when did they arrive? This question is more easily answered with respect to plants than animals, which were not monitored until 25 years after the eruption. No plant life was detected on visits in October 1883, 2 months after the eruption (when the summit of Rakata was reached), or May 1884, but by September 1884 a few blades of grass were noted. In 1886, 3 years post-eruption (E + 3), Treub (1888) noted a film of cyanobacteria, which he believed acted as a substrate for the ferns dominating Rakata's interior, but no fern cover was evident on the other two islands. Twenty-four species of higher plants (sperma-tophytes and pteridophytes (11 species of ferns)) had become established, including two grasses (*Neyraudia madagascariensis* and *Pennisetum macrostachium*), and total destruction of Krakatau's flora was indicated by the absence of resprouting trees or clumps of *Saccharum*. This contrasts with the situation on Volcano Island (above), where, in the same period, seven times as many species (176) had colonized, almost half thought to have been dispersed by birds, 31 (18%) by wind and only 13 (7%) by water. Also in contrast with Volcano's 3-year vegetation, the ash emplaced on the Krakataus was both hot and deep, and the water gap between source biota and target island was much greater than it was in the case of Volcano.

Three years later (E + 6) the number of vascular plant species had risen to 64, and in a further 8 years (E + 14) savanna grassland (with *Saccharum spontaneum*, *Imperata cylindrica* and, less prominently, *Spinifex littoreus*), with patches of *Casuarina* woodland and a few *Ficus* and other trees, had replaced the ferns in Rakata's lowlands.

Casuarina equisetifolia is a dioecious dicotyledonous tree that can reach 40–50 m in height. It is the most important non-leguminous tree to fix atmospheric nitrogen. This occurs within the root nodules, where the filamentous soil bacterium *Frankia* produces a nitrogenase that catalyses the reduction of atmos-pheric nitrogen to organic nitrogen within the plant. In sandy soils on the Cape Verde Islands, the species produced an annual increment of about 58 kg of nitrogen per hectare (Dommergues 1966). The root systems are also usually associated with either ectomycorrhizal or endomycorrhizal soil fungi which facilitate uptake of phosphate and in some cases water. Microorganisms may also produce proteoid roots, a network of fine surface roots that is very effi-cient at taking up nutrients. The *Casuarina* host plant benefits from all these associations, which, together with its rapid growth, make it an excellent colonizer of primary substrates. *Casuarina equisetifolia* is used as a plantation tree in many tropical countries and grows to a height of up to 4 m in 1 year, 10 m in 5 years, 24 m in 10 years, and 28 m in 25 years. It does not survive high-intensity fires and does not regenerate after light fires. It is important in some beach successions and it has long been known that the presence of *Casuarina* improves sandy soils so much that regeneration of other plants is facilitated (Narashimhan 1918). Torrey (1983) noted that species of *Casuarina* are pioneers in a number of habitats, and precede, and by their nitrogen-fixing abilities make possible, the establishment of forested stands. Torrey also mentioned that well-nodulated new seedlings and young trees of *C. equisetifolia* were successful primary colonists (with sedges) of an area of new volcanic ash at Punha, Hawaii Island. Silvester (1977) summarized the evidence for the importance of the species in the revegetation of Krakatau. The litter is noxious in some way and inhibits germination or seedling growth, so that nothing (including its own seedlings) grows in its shade. *Casuarina equisetifolia* itself is intolerant of shade and when other pioneer trees become established it does not regenerate beneath them.

By 1897 (E + 14) a beach association had developed, dominated by the creepers *Ipomoea pes-caprae* and *Canavalia rosea* and including the grass *Ischaemum muticum*, often with a *Terminalia–Barringtonia* woodland of widely distributed sea-dispersed trees and shrubs behind it. By 1919 (E + 36) mixed, species-poor secondary forest, initially with heliophilous small pioneer trees such as some *Ficus* species and *Macaranga tenarius* dominant, began to replace the savanna grassland and *Casuarina* woodland. In the 1930s (E + 50) a *Neonauclea* forest developed, first in the uplands, and the forest canopy began to close. On Sertung and Panjang the dominant trees of the closing forest were two other species, *Timonius compressicaulis* and *Dysoxylum gaudichaudianum*, which, although present on Rakata, did not dominate there.

However, between 1934 and 1979 only one partial floristic survey of the Krakataus was undertaken (Whittaker *et al.* 1989), despite the advocacy (noted below) for regular surveys to be instituted. These early data on plants were discussed extensively by Whittaker and his colleagues (Figs. 8.2, 8.3). Rakata, as a largely uninterrupted primary succession for well over a century, is of particular interest in assessing modes of plant dispersal, and indicating the increasing importance of animal dispersal of propagules once more simple communities have become established from sea- or wind-borne propagules (see Chapter 14, p. 205).

The first animal reported from the Krakataus after the eruption was a single small spider found by Cotteau (1885) about 9 months later. A larger variety of invertebrates (not identified to species level) and a 'gigantic lizard' (almost certainly the monitor *Varanus salvator*, an excellent swimmer) were reported by Selenka and Selenka (1905) during their observations in 1889. By 1908 the Krakataus had already received over 200 species of animals. Indeed, this post-

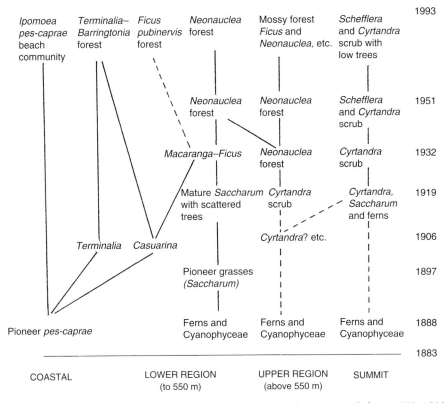

Figure 8.2 The main successional stages of vegetation development on Rakata, 1883–1988. (After Whittaker *et al.* 1989, Bush *et al.* 1992.)

eruption animal survey by E. Jacobson (1909) yielded this number in only 3 days on the islands. Most of Jacobson's survey time was passed on Rakata, with a few hours each on Panjang and Sertung. Nearly all of the animals seen were insects, but the limited spectrum of collection methods implies that many taxa could also have been overlooked. No mammals were found, with the absence of bats specifically noted, but 13 species of non-migrant land birds were reported. In the already well-developed *Terminalia* forest the fruit-eating pink-necked green pigeon (*Treron verans*) and the ground-frequenting emerald dove were seen, and several other birds were insectivores.

Direct observations of animal arrivals in those early days were not recorded, but some of the insects (such as the large carpenter bee *Xylocopa latipes*) reported by Jacobson fly strongly and were seen by our group traversing the open water between islands of the archipelago. Others, such as early-arriving termites, are more likely to have arrived by rafting (p. 209), and the common house gecko (*Hemidactylus frenatus*) by human agency. Generally, though, the paucity of animal monitoring in the few decades post-1883 leaves substantial gaps in our knowledge of community development.

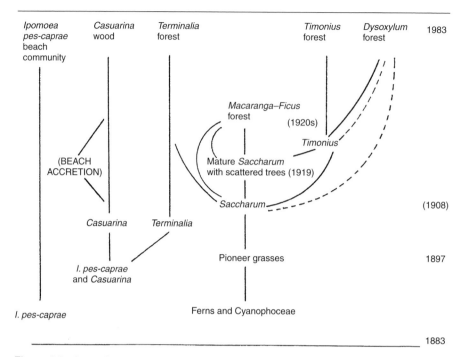

Figure 8.3 The main successional stages of vegetation development on Sertung and Panjang, 1883–1988. (After Whittaker *et al.* 1989, Bush *et al.* 1992.)

Not until 1919, in an expedition associated with the Netherlands Indies Natural Science Congress (held in Jakarta, October 1919), was further serious attempt made to appraise the diversity of animals living on the Krakataus. At this time, the first bat (a dog-faced bat, *Cynopterus brachyotis angustatus*) was found, on Rakata and Sertung. An additional dozen species of non-migratory land birds also occurred whch, allowing for non-rediscovery of two of the species found by Jacobson (a kingfisher and the sooty-headed bulbul, *Pycnonotus aurigaster*) brought the total to 23 recorded species. Although some of the new species might have been overlooked previously, most were likely to be new arrivals.

An important outcome of the Jakarta conference in 1919 was the decision to henceforth survey the Krakatau biota at more regular intervals, leading to the participation of Docters van Leeuwen (Botany) and K. W. Dammerman (Zoology) which proved so pivotal in setting the firmest possible base-line for later studies, from 1920 onward. Thus, Dammerman visited the islands six times from 1919 to 1922, working mainly on Rakata and Sertung. He recorded 777 species of animals over that period: 621 on Rakata and 335 on Sertung (Dammerman 1948), to set a firm comparative foundation for later appraisal. Dammerman's

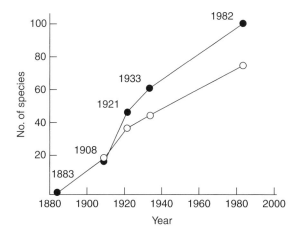

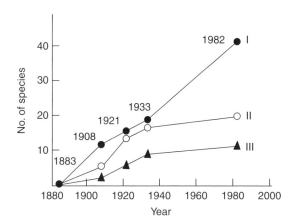

Figure 8.4 Colonization of the Krakatau Islands by aculeate Hymenoptera. Upper: species accumulation and numbers of aculeate Hymenoptera species, other than ants, on Rakata and Sertung combined (filled circles, cumulative numbers; open circles, actual numbers on survey occasions). Lower: species accumulation curves for segregates of aculeate Hymenoptera on Rakata and Sertung combined: I, Apoidea, excluding *Apis*, and Sphecidae; II, Pompilidae, Mutillidae, Scoliidae and Evaniidae; III, social bees (*Apis*) and wasps (Vespidae). (After Yamane *et al.* 1992.)

second series of expeditions (1932–34) revealed some groups of animals apparently to decline from 1920; thus fulgorid bug species found decreased from 19 to eight, and crickets from seven to two. Because the same observer was involved, rather than different workers with different perception skills, these declines may be genuine rather than the consequence of sampling vagary. Notable changes amongst the vertebrates included advent of a second species of snake (possibly the paradise tree snake, *Chrysopelea paradisi*) and a third gecko (*Corymbotus platyurus*).

Examples of faunal change, extended to include surveys by my group and others to the early 1990s, are shown in Figs. 8.4–8.8. The land birds of Rakata are perhaps the most thoroughly appraised group, and Table 8.1 summarizes the extensive information available on these.

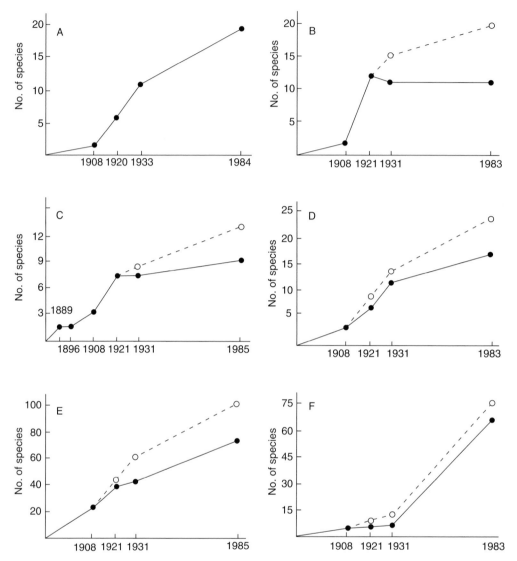

Figure 8.5 Colonization of the Krakatau Islands by selected groups of animals: filled circles, actual numbers; open circles, accumulative species numbers. A, land molluscs; B, Odonata (dragonflies); C, reptiles; D, Blattodea (cockroaches); E, Hymenoptera, Formicidae (ants); F, Hymenoptera, Braconidae. (Based on Thornton 1991.)

During the 1930s, much of the interest in the Krakataus shifted to accommodate the emergence of Anak Krakatau, treated by biogeographers as a chance to study biotic succession as a 'model within a model' system of community development, for which the organisms already present on the surrounding older islands are likely to constitute the predominant sources of colonists. By

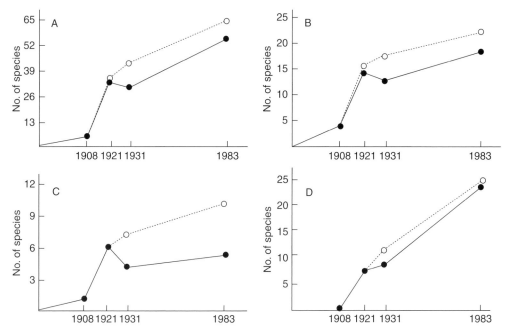

Figure 8.6 Colonization of the Krakatau Islands by butterflies: filled circles, actual numbers of species; open circles, accumulative numbers of species. A, all butterflies; B, Nymphalidae; C, Hesperiidae; D, Lycaenidae. (Based on Thornton 1991.)

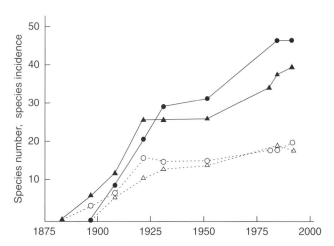

Figure 8.7 Colonization of the Krakatau Islands by species of figs and flying frugivores. Species incidence curves (filled symbols) are plots of the sum of number of islands (Rakata, Panjang, Sertung) occupied by each species for each survey date; filled circles, frugivore species incidence; filled triangles, fig species incidence; open circles, fig species numbers; open triangles, frugivore species number. (After Thornton *et al.* 1996.)

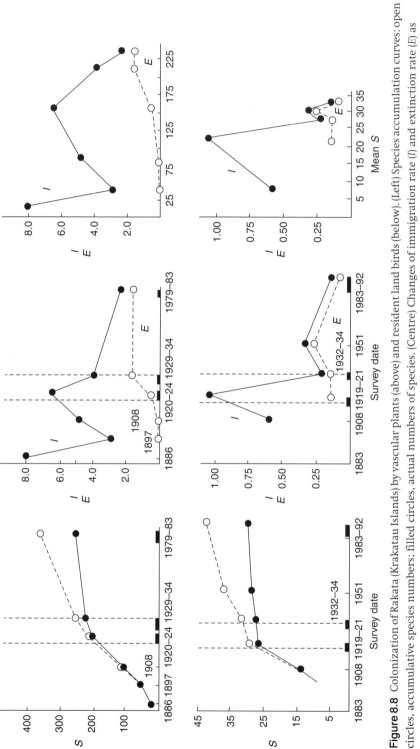

Figure 8.8 Colonization of Rakata (Krakatau Islands) by vascular plants (above) and resident land birds (below). (Left) Species accumulation curves: open circles, accumulative species numbers; filled circles, actual numbers of species. (Centre) Changes of immigration rate (*I*) and extinction rate (*E*) as species per year for intervening intervals denoted. (Right) Changes in *I* and *E* with average number of species (mean *S*) in inter-survey intervals. Vertical dotted lines (left, centre) represent main period of forest formation on Rakata; dark basal horizontal bars are survey dates with data combined over each such period. (After Thornton *et al.* 1993.)

Table 8.1 *Resident land birds of Rakata, 1908–92, based on assessment of records on basis of minimum turnover*

Inter-survey period	I	II	III	IV	V
Survey dates		1908	1919–21	1932–34	1951	1983–93
Inter-survey interval (years)	25	13	13	17	41	
Actual number of species		15	27	28	29	31
Cumulative number of species		15	29	32	38	43
Gains	15	14	3	6	6[a]	
Losses	–	2	2	5	4	
Immigration rate[b]	0.60	1.08	0.23	0.35	0.15	
Extinction rate[b]	–	0.15	0.15	0.29	0.10	

[a] One immigrant, *Ducula bicolor*, colonizes opportunistically for short periods.
[b] Given in species per year.
Source: Thornton *et al.* (1993).

contrast, the post-1883 biota arriving on Rakata, Sertung and Panjang must be from considerably farther away, the mainland of Java or Sumatra, with the possibility of 'stepping stones', particularly the more northerly island of Sebesi, discussed later in this book.

PART IV

Assembly of biotas on new islands

Starting points

One problem with studies of recovery after presumed extirpation of a biota is the doubt that almost always exists about the completeness of the eradication of life. For example, in the case of Volcano Island (Taal Volcano) it is known that a small remnant of vegetation survived the eruption, and doubts concerning the complete eradication of both animals and plants have been expressed concerning the eruptions of Thera (Schmalfuss and Schawaller 1984) and Krakatau (e.g. Backer 1929, but see Docters van Leeuwen 1936 and Dammerman 1948). In his children's fantasy *The Water Babies*, Charles Kingsley (1915: 61, 62) pointed out that 'no one has a right to say that no water babies exist, till they have seen no water babies existing; which is quite a different thing, mind, from not seeing water babies . . .'. Absence of evidence is not necessarily evidence of absence, and this is certainly true for seed plants, which may survive, unrecorded, in seed banks in the soil (Whittaker *et al.* 1995), as well as for animal groups whose mobility and cryptic or nocturnal habits may mean that they escape detection even in specialist surveys.

On newly created land there can be no question of 'not seeing water babies'. As far as land organisms are concerned the substrate is new, clearly virgin, abiotic, and, if studied soon enough (as on Surtsey), the beginnings of a primary xerosere (terrestrial succession) can be observed.

Marine cays are such newly created islands. Although they are small and low, and occur on continental shelves, sand cays or coral keys are young oceanic islands in the biological sense. They were formed by the deposition of sand or coral, without a land biota, and their land animals and plants have reached them over sea water.

Volcanic islands that emerge from the sea as hot steaming masses of ash or lava are also newly formed islands that must be completely devoid of land life when they are formed, and they provide all but one of the other case studies to be examined in this section. There are several such islands available for study, some of which emerged from the sea, others from freshwater lakes. Some of these are little known and have not been studied. In several cases the new island emerged from a submerged caldera, when the volcanic activity that created the caldera resumed many years later. Examples of emergent marine islands are Nea and Palea Kameni in the caldera of Thera, Santorini archipelago, Mediterranean; Rakaia (Vulcan) Island, in the Rabaul caldera, New Britain, which became joined to the mainland in an eruption 59 years after its emergence; Anak Krakatau in Krakatau's caldera, Sunda Strait, Indonesia; Surtsey and the lesser-known Thingvallavatn on the mid-Atlantic ridge, off Iceland; Ponta los Campelinhos in the Azores, which emerged during an eruption in 1957 and later merged with the main island, Faial; and Tuluman, south of Manus Island, in the Admiralty Group of the Bismarck archipelago north of New Guinea. The group of small islands making up the beautiful little Banda archipelago in the Moluccas are almost certainly the remnants of a former huge 'Banda Volcano', which erupted explosively in prehistoric times. Gunung Api, the active 'Fire Mountain' island near the middle of the group may have emerged from the partially submarine caldera resulting from the massive eruption, but there are no estimated dates for this event.

There are at least three emergent islands within freshwater calderas or crater lakes – Motmot, in the caldera lake of Long Island (off New Guinea); Volcano Island, in the caldera lake of Taal Volcano (Luzon, Philippines); and a small island formed in the crater lake of Deception Island (off Antarctica). Several of these islands are emergent within the calderas of volcanoes that are themselves recovering from volcanic effects.

Finally, there is at least one case of an emergent island of water, as it were – a newly created lake within the caldera of Long Island – Lake Wisdom. Not all these cases have been studied by biologists, and those that have been studied have received different amounts of attention. The best studied, roughly in order, are Surtsey, Anak Krakatau, Motmot, Lake Wisdom, Tuluman, Vulcan and Volcano (in the last two, only the plants were surveyed). The Kamenis emerged in about 200 BC, well before ecologists were available to study the newly emerged islands, although the Santorini Group is now being studied by a largely German team of biologists (Schmalfuss *et al.* (1981 and later papers)). The Banda emergence, if indeed Gunung Api is an emergent island, was prehistoric.

The emergent marine islands on which the most intensive, long-term studies have been carried out are the active island volcano Anak Krakatau (Krakatau's Child) in Sunda Strait, Indonesia, and Surtsey, off Iceland in the North Atlantic. Because of the long-term studies carried out on them, each merits a chapter to itself.

CHAPTER NINE

Lake Wisdom: a new island of fresh water

It may seem rather odd to include a lake in a chapter concerned with emergent islands, but of course for freshwater organisms a newly created lake is a pristine island of water surrounded by land, inhospitable areas which must be overcome for colonization to occur.

Lake Wisdom was discovered by Westerners in 1928, when an official party climbing the island reached about 400 m and were astonished to see, some 200 m below them, the waters of a huge lake. It was named Lake Wisdom, not because the drinking of its waters (which are clear and refreshing) leads to the getting of wisdom, but to honour Brigadier General E. A. Wisdom, Administrator of New Guinea from 1921 to 1933. Inhabitants of Long Island's sea coast, then numbering about 300 immigrants from the Siassi Islands (7 years earlier Long Island had been reported to be uninhabited), said that in the highlands there lived people whom they had never seen, but whom they blamed for the disappearance of their women from time to time.

Long Island is basically a large andesite volcano, whose explosive eruption, probably in about 1645 (see Chapter 7), resulted in the formation of a huge central caldera which takes up a third of the island's area. Over the years Lake Wisdom was formed as rain run-off filled the caldera. The lake now covers an area of about 95 km^2, has a maximum diameter of about 13 km, and is surrounded by the ring of Long Island which averages about 8 km from lake shore to ocean shore. The lake surface is some 190 m above sea level and the greatest depth plumbed is over 360 m – more than 170 m below sea level – yet the water at that depth is not at all brackish; the lake is well sealed from the ocean. The caldera walls consist of cliffs rising some 140 to 260 m above the lake surface; in the north the rim reaches a height of about 450 m above sea level, so the caldera has a maximum depth of some 620 m. Two mountains rise from the caldera plateau, Mt Réaumur (1280 m) in the north and Cerisy Peak (1112 m) in the south (Fig. 7.2, p. 98). Except for two shallow bays on the western and eastern shores, the caldera cliffs continue steeply into the lake, without a marginal shallow shore zone.

Knowledge of the limnology of Lake Wisdom is entirely due to the researches of Dr Eldon Ball and his colleagues. Overcoming great difficulties, they studied the lake on a number of visits in the 1970s and 1980s (Bassot and Ball 1972, Ball and

Table 9.1 *Summary of the biotic differences between Lake Dakataua and Lake Wisdom*

	Dakataua	Wisdom
Beds of rooted *Chara* and *Najas*	present	absent
Mollusc species	6	4
Cladoceran species	7	2
Copepod species	1	0
Aquatic Hemiptera species	8	5
Dragonfly species	9	3
Chironomid species	7	10? +
Frog species	0	2
Freshwater fish species	0	0

Glucksman 1978, 1980, 1981, Specht *et al.* 1982). Ball and Glucksman compared its limnology with that of the nearest body of fresh water of comparable size, also a caldera lake – Lake Dakataua in New Britain, at the same latitude as Lake Wisdom but 285 km to the east of Long Island. Somewhat smaller than Lake Wisdom (about half its area, a sixth of its volume and a third of its depth), with a more complex bottom topography, it is also older, having been formed at least 1000 years ago. Lake Wisdom was found to be unusual in a number of ways (Table 9.1).

First, in spite of its considerable depth of more than 360 m, there is very little of the expected stratification in temperature, acidity or oxygen saturation, and freshwater molluscs were found living on the deepest part of the bottom plumbed (see above, Chapter 7). Both lakes show a thermocline between 10 and 25 m but in Lake Dakataua temperature then declines more rapidly with depth, and there is another thermocline at 40–45 m. Lake Wisdom's surface temperature (28 °C) (it is at 190 m altitude) is lower than that of Lake Dakataua (31 °C) (at 76 m altitude), and there is no obvious second thermocline. Indeed, at a depth of 50 m the temperature is only 1–2 °C below that of the surface water and the temperature at a depth of 360 m was estimated to be more than 22 °C. Corresponding to the relative lack of temperature stratification, whereas in Lake Dakataua acidity increased consistently with depth, Lake Wisdom showed no clear relation between pH and depth. In addition, whereas in Lake Dakataua there were sharp gradients in oxygen saturation at 22 and 40–45 m and no measurable oxygen below 80 m, oxygen saturation in Lake Wisdom was relatively high right down to the lowest depth plumbed, 360 m. Living organisms were common in Lake Dakataua down to 20 m but absent below this depth, and bottom samples taken from depths greater than 60 m gave off a smell of hydrogen sulphide. In contrast, samples from 360 m in Lake Wisdom had no detectable smell of hydrogen sulphide and living molluscs and chironomid larvae were present at that depth.

Ball and Glucksman thought that the generally poor stratification of Lake Wisdom, unusual in such a deep lake, might result partly from persistent mechanical and convectional disturbance. Unlike Lake Dakataua, which is almost completely divided into two just below the surface, Lake Wisdom is a single basin, and any mechanically or convectionally induced currents involve the entire volume of water. Lake Wisdom's steep caldera walls rise some 140 to 260 m above the lake surface but do not shelter it from persistent south-easterly (June to October) or north-westerly (October to May) winds blowing at 350 to 450 m above sea level. There is a good fetch across the 13-km-diameter lake, and waves over a metre high are common; indeed, parts of the lake's western shoreline are eroding at a rate of about 2 m per year (60 m in 29 years: Ball and Glucksman 1978: 458). Jared Diamond (Diamond 1974b) was unable to cross to the lake islet Motmot from the western shore in an outrigger canoe in August 1972, and Ball and Glucksman were only able to work on Wisdom's limnology by making their visits at a time when the wind was changing direction (October–November).

Volcanic heat effects may also prevent stratification. The temperature of sur-face water at the margins of Motmot was measured by Ball and Glucksman in 1969–72 and found to vary between 29 and 90 °C from place to place, and Osborne and Murphy in 1988 found that it varied from 30 to 80 °C (Osborne and Murphy 1989). The heat focus at the Motmot vent (which probably connects with the same magma chamber involved in the explosive eruption of Long Island) must drive convection currents, and these are probably more important than wave action in producing a circulation of water throughout the lake's volume.

The second unusual feature of Lake Wisdom is the simplicity of its biota. The extensive beds of rooted aquatic plants, *Najas tenuifolia* and *Chara* species, found in the shallow areas of Lake Dakataua have no counterpart in Lake Wisdom. In most groups of animals, also, Wisdom is less diverse than Dakataua: four species of molluscs compared to six; two planktonic cladocerans and no copepods com-pared to seven cladocerans and one copepod; five aquatic bugs compared to eight; and three dragonflies compared to nine. The exceptions are chironomid midges and frogs (which may not have been completely sampled): at least ten species of midges occur in Wisdom compared to seven in Dakataua, and Wisdom has two frogs but none were found in Dakataua. There are no fish in either lake.

Why Lake Wisdom has no fish

John Asong (1984) related a legend from the Siassi Islands which explains the absence of fish in Lake Wisdom. A woman and her two sons lived on Long Island below the volcano, many years ago. The only available meat on the island was pig, which they hunted, and the family ate pork every day. As they were working in the garden one day, a wild pig ran to them and said, 'Don't be afraid, I have something to tell you. If you agree to stop killing us for a while I promise something good will happen to you.' The wild pig then ran away, leaving them wondering about what it had said for the rest of the day.

In the evening the woman went for a walk and, climbing a hill, looked down on a lake which she had never seen before. It was full of tuna fish jumping about in the clear water. She walked into the lake a short way and found that the fish were so tame that she could pull them out by their tails. She caught several and took them home to cook, but would not tell her sons where the new food had come from. For weeks she would go off alone in late afternoons, returning with fish for supper. Finallly, one afternoon the boys followed their mother and, from behind bushes, watched her go into the lake and pull out tuna by their tails. After she left they ran to the lake and began pulling out scores of tuna. They didn't want to tell their mother that they had followed her, so they decided not to take them home (there were too many to carry in any case) and left large numbers lying on the lake shore.

Next afternoon when the woman went to the lake she discovered that all the tuna fish had gone. There were no fish in the lake. They had found a way to escape to the sea. Ever since that day, tuna can only be found in the sea, and Lake Wisdom has been devoid of fish.

Three possible main reasons for the very simple and impoverished nature of Lake Wisdom's biota have been suggested by Eldon Ball and colleagues, none of which involves the escape of an originally marine fauna to the sea (see box). First, Lake Wisdom is extremely isolated, both geographically and biologically. The nearest body of fresh water of comparable size (Lake Dakataua) is 285 km distant and Umboi's smaller lakes are some 80 km away, both to the east. Freshwater propagules have a double hurdle to overcome if they are to reach Lake Wisdom – the barrier of sea water isolating Long Island, and Long Island's ring of land surrounding the lake itself. A second possible reason for the biological simplicity is that the lake is young. It almost certainly developed by rain-fill of the new caldera that was formed by successive slumping following the explosive eruption in about 1645. So there has been little time, about three and a half centuries, for its colonization from other freshwater sources. Even copepods, found in fresh waters all over the world, evidently have not yet colonized. The recency of colonization by its only known sponge species, *Spongilla alba*, is indicated by the presence of gemmules, internal asexual dormant buds that become active on the death of the parent sponge. Evidently freshwater sponges often lose the ability to form gemmules when they have been in deep lakes with small fluctuations of level for a few thousands of years (Ball and Glucksman 1980). Finally, the lake's generally steep-sided shores and relative lack of shallows perhaps partly explain the surprising absence of aquatic vacular plants (and thus an important limnic habitat for other organisms), although I agree with Ball and Glucksman (1978: 467) that 'the northwest corner seems to be ideally suited to them with its extensive areas of shallow water over a mud bottom'.

CHAPTER TEN

New islands in the sea

Sand cays in the Coral Sea and on the Puerto Rico Bank

In their studies of community assembly, Heatwole (1971, 1981) and Heatwole and Levins (1972a, 1973) have taken sand cays as their study units, to good effect.

Heatwole (1971) examined 12 bare cays of the Great Barrier Reef and adjacent areas of the Coral Sea, north-east of Australia. He found that although the cays were at least 200 km from the mainland and quite devoid of vegetation, all of them contained terrestrial invertebrate animals.

Heatwole was able to record a number of instances of dispersal. Only one moth (a microlepidopteran) was found on these plantless cays, clearly an unestablished immigrant; an individual of the same species was seen at sea 3 km from the cay. On one cay the empty pupal cases of a large fly testified to another colonization attempt. On another, only a single centipede was found, and this on the beach beneath a washed-up coconut with a crack in its husk, in which the centipede may have survived the sea journey.

The sequence of colonization followed on the bare cays was the same as that outlined by Dammerman on Krakatau (Chapter 8), with scavengers and detritivores exploiting ocean-derived food and forming the base of the islands' own trophic webs. There was a high rate of turnover and the numbers of species present on the islands were related to island area, height and use by sea birds, which acted as conduits of energy from the ocean to the island by feeding on fish and contributing guano, their dead bodies as carrion, and occasionally regurgitated fish to the terrestrial island system.

Palea Kameni and Nea Kameni (Santorini), b. 197 BC

Since the great Minoan eruption of Thera three and a half millennia ago (Chapter 6), volcanic activity on the Santorini archipelago has been localized largely to an area within the new, enlarged caldera, and along a north-east–south-west trend line. The only exception has been activity in the sea 6 km to the north-east of Thera, near the present Kolumbo Reef. Post-Minoan activity in the Santorini group has been detailed and mapped by Fytikas et al. (1990), but according to Krafft (1991) the first person to have written about volcanic activity in the Santorini islands was the Greek historian Poseidonius, whose writings were transcribed by the Greek geographer Strabo.

Poseidonius recorded that an island, Hiera, rose up 'as if on springs' from the submerged caldera of Thera in 197 BC (Krafft 1991). This activity, which was both explosive and effusive (i.e. producing lava flows), quickly built up a largely pyroclastic cone known as Iera or Hiera. According to Aurelius Victor in *Historia Romana*, nearly two and a half centuries later, after Iera had disappeared, in AD 46 and 47, volcanic activity about 3 km south-west of Iera's position formed an island, Thia, about 1 km long and 500 m wide.

Very much later, in AD 1570, there was a highly explosive eruption a few hundred metres to the north-east of Thia. This was followed by a lava flow which fused with Thia, thus slightly enlarging the island to form the island Palea Kameni ('old burnt island'). Between 1570 and 1573, further volcanic activity within the caldera formed another island, Mikra Kameni ('small burnt isand'), some 2 km to the north-east of Palea Kameni.

Some eight decades later, in 1650, and following an intense series of earthquakes over a period of 2 years, a cone that had been built up from a base 280 m below sea level broke the sea surface some 6 km to the north-east of Thera. There were strong explosions, with pumice and ash production, and a tsunami caused serious damage to Thera and other islands within a 150-km radius. The sulphur dioxide emitted by this Kolumbo volcano drifted to Thera, killing 50 people and more than 1000 animals. On Thera, silver kept in wooden boxes turned black, and people on Naxos and Crete became ill as a result of sulphur pollution. The small cone which emerged from the sea was soon eroded and is now 20 m below the surface, forming Kolumbo Reef.

From 1707 to 1711 further outpourings of lava from the active centre in Thera's caldera built another emergent island, Nea Kameni ('new burnt island'), close to the southwest shore of Mikra Kameni. It was partly as a result of observing the birth of Nea Kameni that the Italian priest Anton-Lazzaro Moro advanced the theory that volcanic activity involves considerable uplift of the ground in the area and that this is how all the Earth's continents, islands and mountains were formed (Krafft 1991). Nea Kameni was tripled in size as a result of a mainly extrusive volcanic episode that extended from 1866 to 1870. Fifty years later, in a largely explosive episode of volcanism lasting from 1925 to 1928, Nea Kameni fused with Mikra Kameni, forming a roughly circular single island about 2 km in diameter – the Nea Kameni of today.

The French geologist Professor Ferdinand Fouqué, of the Collège de France, was sent to Santorini immediately after the 1866 eruption of Nea Kameni. His investigations resulted in a classic geological, petrographic and historical study (Fouqué 1879, 1998). He calculated that the Thera caldera could not have resulted simply from the destruction of a pre-existing volcanic cone, for there was insufficient material present as rock fragments to account for the missing volume, which he therefore concluded must have slumped below the sea. Fouqué's study of the stratigraphy of the caldera wall revealed archaeological remains dating from 2000 BC under a 30-m-thick layer of pumice, proof that the island had in fact collapsed after an enormous explosive eruption, and laying to rest the previously accepted 'theory of uplifted craters', which, he observed with obvious enjoyment, 'can henceforth be considered only as one of those noble wrecks that litter the path of science'.

The caldera was indeed a collapse caldera, formed when the submarine magma chamber became depleted and could no longer support its own roof, which collapsed into the empty chamber as the northern half of Thera was broken into three separate islands surrounding a 500-m-deep bay.

Fouqué's famous visit to Santorini immediately after the 1866 eruption (Chapter 6) was made when, following Moro's ideas, Nea Kameni was still considered to be one of the world's best examples of sudden colossal uplifting. The visit was extremely productive in terms of volcanological advances. As well as disproving Moro's theory by showing that the Thera caldera was formed by collapse rather than uplift, Fouqué (1879) also plotted the development of Nea Kameni. He described the growth of its viscous lava domes, concluding that the viscosity of the lava determines the shape of volcanic structures. He found two types of feldspar crystals in the island's lava, and was the first to point out the phenomenon of magma mixing deep in the Earth, a matter destined to receive much more attention over a century later. Fouqué also showed that flames accompanying the lava flows were a burning mixture of hydrogen and methane, and he was the first volcanologist to realize that hydrogen originates deep within the Earth (Krafft 1991).

Strong intermediate-depth seismic shocks in the Santorini archipelago in 1923, 1926, 1937 and 1948 were followed by eruptions in 1925, 1928, 1939 and 1950 respectively (Blot 1978). There was effusive–explosive volcanic activity on Nea Kameni from August 1925 to January 1926, when the eruption column reached a maximum height of 3200 m. After 4 months of inactivity, water–magma interaction produced limited pyroclastic flows in May, and between then and January 1928 there were four phreatic explosions followed by some extrusive activity which built up a lava dome. A phreatic explosion is the result of magma overheating water contained in the lavas, new magma appearing later in the resulting crater. Between August 1939 and July 1940 there were five phreatic explosions, all but the last being followed by a lava flow producing a dome. For 3 weeks in early 1950 phreatic explosions were followed by lava flows having the same chemical composition as those of the last 500 years. This was the last volcanic activity recorded in the Santorini group at the time of writing (2006). The interval between eruptive episodes has been declining, in general; since 197 BC repose times have been 243, 679,

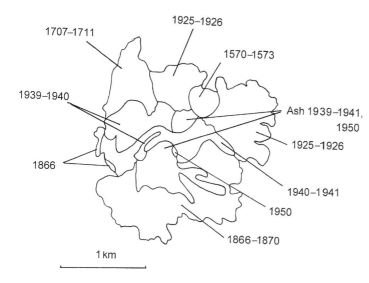

Figure 10.1 Santorini: Nea Kameni to indicate the volcanic mosaic of habitats produced by volcanic activities at different times (shown) (cf. Anak Krakatau, Fig. 11.3). (After Schmalfuss *et al.* 1981.)

844, 134, 155, 55, 11, 9, and 49+. The highly explosive eruptions of 197 BC, AD 726 and AD 1650, however, are separated by almost equal intervals of just over nine centuries.

The marine-emergent Kameni islands were first surveyed by biologists in 1819, when nine species of plants were recorded, and by 1881 there were 20 species. By 1899 there were 49 plant species on Nea Kameni alone (29 of them composites and grasses), and in 1911, 75. After the eruptions of 1925 and 1926, however, the number of species fell to 26, largely through the loss of annuals. Nevertheless, by 1967 the flora had been rebuilt to 69 species, and by 1984, to 116, immigration of species having outstripped extinctions incurred during the eruptions in 1939–41 and 1950 (Diapoulis 1971, Hansen 1971, Raus 1986, Sipman and Raus 1995). I have been unable to find any more recent assessment of the flora's dispersal-mode spectrum, although Raus (1988) gave additional comments on this. The mosaic nature of the flora was demonstrated well by Raus's (1986) maps showing the limitation of particular plant species to one (or few) of the different-aged substrates shown in Fig. 10.1.

The tenebrionid beetle fauna of the Santorini archipelago has been documented by Grimm (1981) and the sizes of the islands' faunas appear to be related to the sizes and ages of the islands. There are 29 species in all. Of the remnants of ancient Stronghyli, 28 species occur on the largest island, Thera, 13 on Therasia, and six on the small remnant Aspronisi. On the relatively young, 2000-year-old Palea Kameni there are ten species, and the much younger 300–500-year-old Nea Kameni evidently carries only one (Grimm 1981). The wider survey of animals on the archipelago that initiated the above (and other) taxonomic papers showed a wide array of taxa collected or observed in

1978 and 1979. These were listed by major taxon by Schmalfuss *et al.* (1981), but many await further analysis.

Tuluman, Admiralty Group, Bismarck Sea, b. 1953
Tuluman's emergence

About 30 km south of Manus Island, the largest island of the Admiralty Group in the northern Bismarck Sea and famous as a field study island of anthropologist Margaret Mead, is a 7-km-long island, Lou, formed of a chain of overlapping volcanic cones of Quaternary age. Volcanic activity was reported in the sea about 1 km south of Lou in 1883, and the 1944 Admiralty Chart (Aus. 054) shows a shoal about 2 km south of the island, possibly at the same site as the 1883 activity. However, no shallows had been reported, nor was there any indication of impending volcanic activity from seismic monitoring, when, in 1953, the first of a series of eruptions occurred.

The island-building eruptions, at eight different eruptive centres, lasted from 1953 to 1957 (Johnson and Smith 1974, Reynolds and Best 1976), and consisted of submarine extrusion as well as explosive and effusive activity at the surface. In 1954, a small island, Tuluman, emerged from the sea. A phase of effusive activity, mostly submarine, lasted for some 20 months, those vents reaching the surface subsequently undergoing explosive phases, with eruption columns of vapour and ash rising up to heights of up to 5 km. The climax was reached in February and March, 1955, when there was an intense period of explosive activity, with an almost continuous series of surface explosions.

Following a year of relative quiescence, a final, major effusive phase began in November 1956 and lasted until 28 January 1957. Two new islands resulted: Tuluman, some 700 m long and about 25 m above sea level (30 m at the crater rim), and a small 90-m-long islet of lava, known as 'cone 3 island', 400 m to the north-west. Lava flows covered the north-eastern two-thirds of Tuluman's surface together with lava bombs, lapilli, pumice and ash, and pyroclastic flow deposits alone covered the remainder. Less significant thermal activity occurred on Tuluman at least up to 1964, but by 1971 this had ceased.

Thirty years after its birth, when the island had a maximum diameter of about 500 m, a height of some 23 m and an area of about 0.28 km^2, the colonization of Tuluman by plants and animals was monitored by the New Guinean biologist Karel Kisokau and his colleagues (Kisokau *et al.* 1984).

Tuluman's colonization

The colonization of Tuluman and Motmot, islands of about the same size and age in the same geographical and climatic area, are in sharp contrast, and this highlights the importance of the nature of the barrier in determining an island's pioneer colonizers. In contrast to Tuluman, Motmot has a triple barrier, of sea (to Long Island) and then land and fresh water (to Motmot). In addition it is further from its putative source (5–7 km from Long Island) than is Tuluman (1–2 km from Lou).

Among plants, the seaborne pioneers of Tuluman, such as *Hibiscus tiliaceus*, *Terminalia catappa*, *Pandanus tectorius*, *Barringtonia asiatica*, *Premna obtusifolia*, *Morinda citrifolia*, *Calophyllum inophyllum* and *Scaevola taccada*, all of which, incidentally, were important pioneers of the emergent marine island Anak Krakatau, are of course excluded from the freshwater island Motmot. Thus, of a combined flora of some 60 species, Tuluman's young flora and that of Motmot have in common only two – the wind-borne *Imperata cylindrica* and the animal-dispersed *Ficus opposita* (p. 201).

CHAPTER ELEVEN

Anak Krakatau, Krakatau's child, b. 1933

Birth and early physical development

In 1930 an island broke the surface of the sea covering the submerged caldera resulting from Krakatau's 1883 eruption (Fig. 3.3, p. 38). It appears to have been built up from a vent that was about halfway between the two main vents of the 1883 eruption, and is almost certainly the product of the same magma chamber. Unfortunately, as was the case after the 1883 Krakatau eruption, the opportunity to study the very early steps in the processes of animal colonization and community assembly was again missed. There have been a number of intensive studies on the island in recent years, however.

The biota of the young active island volcano was evidently devastated by the eruptions of 1952–53 (van Borssum Waalkes 1954) and there have been several severe eruptions since. Such events have provided further opportunities of monitoring the early stages of the assembly and development of a tropical forest island ecosystem on the island. In this case the colonists presumably originated on Anak Krakatau's three older companion islands, themselves in process of recovery from the 1883 eruption. Since a nested pair of natural experiments in colonization was now in progress at the same site, the experiments having started several decades apart, a number of comparative studies became possible (Thornton *et al.* 1992). Anak Krakatau's continuing volcanic activity, which had no counterpart on Rakata (or on Surtsey), has complicated comparisons of biotic assembly by imposing periodic checks (to varying extents) on the successional processes occurring on the island, but it has also provided replicate sequences of succession from zero or near-zero base-lines. The various initial stages of plant colonization that have occurred on the island following its periodic outbursts (Partomihardjo *et al.* 1992) may be compared with one another and with that seen on Rakata after 1883 (Thornton *et al.* 1992), as well as with the prisere monitored so meticulously on Surtsey (Fridriksson and Magnússon 1992). Unlike either Krakatau in 1883 or Surtsey, however, Anak Krakatau is only 2–3 km from an archipelagic biota that itself was selected for good colonizing ability.

Table 11.1 *Diversity of seedlings and juveniles of maritime plants recorded on Anak Krakatau during the east and west monsoon seasons*

	East	West
Total individuals	20	17
Total species	12	11
Total genera	12	11
Total families	11	10

Source: After Partomihardjo *et al.* (1993).

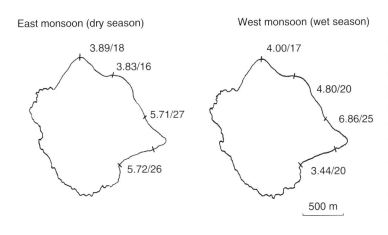

East monsoon (dry season)

West monsoon (wet season)

3.89/18

3.83/16

5.71/27

5.72/26

4.00/17

4.80/20

6.86/25

3.44/20

500 m

Figure 11.1 Incidence of floral propagules on beaches of Anak Krakatau in the two seasons surveyed. Average number of species per transect and total number of species on each beach are shown above and below the line, respectively. (After Partomihardjo *et al.* 1993.)

Rate of establishment of biota

There is good evidence that only a fraction of the propagules that have reached Anak Krakatau have become established, in spite of the fact that most of them probably originated from the neighbouring islands and would thus be components of a subset of the mainland biota that has been selected for good colonizing ability. We have been able to monitor the arrival on Anak Krakatau of propagules and individuals of many plant and animal species since our detailed studies on the island began in 1984 (Table 11.1).

The dispersal of seaborne plant propagules to the island was studied over two different monsoon seasons in 1990–91 (Partomihardjo *et al.* 1993). Stranded propagules of 66 plant species were found on the island's beaches (Figs. 11.1, 11.2). Disseminules of 30 of these species could have been derived from plants growing on the Krakatau islands themselves, mostly from the 'core' of coastal species present on all the islands, but the others must have originated from outside the archipelago, largely from Sumatra, to the north. There are also

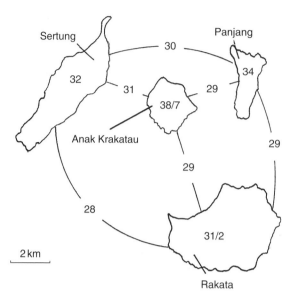

Figure 11.2 Numbers of plant species whose propagules were found on beaches of Anak Krakatau. Numbers on lines between islands are species in common between those islands; numbers below lines for particular islands show the species found only on that island, within the upper or main figure of the total island complement. (After Partomihardjo *et al.* 1993.)

records of the arrival of individuals of some 54 species of animals, 25 of them vertebrates, since 1984 (Thornton 1996a). Fourteen of these (a white-eye (*Zosterops* sp.), leaf-warbler (*Phylloscopus* sp.), house crow (*Corvus splendens*), tiger shrike (*Lanius tigrinus*), serpent-eagle (*Spilornis cheela*), two species of munias, an otter (*Lutra ?lutra*), rousette fruit bat (*Rousettus leschenaultii*), tomb bat (*Taphozous longimanus*), and four butterfly species) were unknown on the other islands and thus were possible new colonists to the archipelago. Over 20 of the 54 arrivals, however, did not become established. Observers were present on the island during this period for a total of only about 4 or 5 months, and it is likely that hundreds of individuals must have arrived, many being species new to the island and thus representing failed colonization attempts.

Pioneers exploiting extrinsic energy sources

Both Bristowe (1931, 1934) and Toxopeus (1950) recorded the presence, at very early stages of Anak Krakatau's biotic succession, of members of a guild of scavengers and detritivores subsisting on organic flotsam, thus indirectly exploiting energy deriving from outside the island, in many cases from the sea itself. The guild included collembolans, chloropid flies, and anthicid and staphylinid beetles. Green turtles, shore birds, crabs, and the omnivorous large monitor, *Varanus salvator*, whose food includes crabs and turtle eggs, have together also acted as food-web links through which marine resources entered the island system.

Another route for extra-island energy was discovered by investigating barren areas of the island, covered in ash and lava. In these areas, but not in vegetated areas, a faunal guild including lycosid spiders, earwigs, a mantis and a (dominant) crepuscular, flightless nemobiine cricket, subsisted on the fallout on to the island of wind-borne invertebrates (New and Thornton 1988, Thornton *et al.* 1988a). Similar guilds of exploiters of allochthonous airborne organic fallout have been discovered in other habitats where primary productivity is low or absent, such as alpine snow fields, oligotrophic lakes and deserts (Edwards 1987), on substrates newly exposed following the retreat of glaciers (Hodkinson *et al.* 2001), and on surfaces covered in lava or ash following volcanic activity on Mount Etna, Mount St Helens, the Canary Islands and Hawaii. Such communities appear to be characteristic of areas with a long history of sustained volcanic activity, and have probably evolved in response to the lack of primary productivity in such areas. Some were treated in detail, with respect to volcanically produced substrates, in Chapter 4.

Energy usually enters an island system as sunlight, and passing through the photosynthetic pathways of green plants to animal herbivores and then carnivores and animal parasites. As we have seen, herbivores generally become established after their plant food has done so, and are generally followed by carnivores, spiders being exceptional early arrivers. The tapping of energy flowing through ecosystems outside the island, through the energy routes outlined above, short-circuits the usual plant–herbivore–carnivore energy pathway and enables the establishment of some animals without the need for the prior establishment of plants.

Successional mosaics

When the current eruptive episode began in November 1992 Anak Krakatau's biota was some four decades old and had been fairly regularly disturbed by volcanic activity, at least twice quite severely (Fig. 11.3). The majority of the island is barren, covered with ash and lava, but along the east coast a considerable community has developed. The ecosystem is at an 'earlier' stage of succession than those of the older islands, which are largely covered in mixed forest, but already includes some 140 species of vascular plants (Partomihardjo *et al.* 1992), 24 non-migrant land birds (Thornton *et al.* 1993), 30 butterflies (New *et al.* 1988, New and Thornton 1992b), three reptiles, and two rats. Seven bat species have been recorded (Schedvin *et al.* 1995) of which two or three may be resident.

The communities on the island's three coastal forelands are not in seral synchrony, however. The north foreland, which from 1979 to 1989 was largely covered in grassland but in 1992 was rapidly changing to *Casuarina* woodland, is at a stage that lags by about 15 years that of the east foreland, where this woodland is well developed and beginning the transition to mixed forest; the north-east foreland is intermediate between the two (Thornton and Walsh 1992). The successional mosaic appears to result from differentially damaging

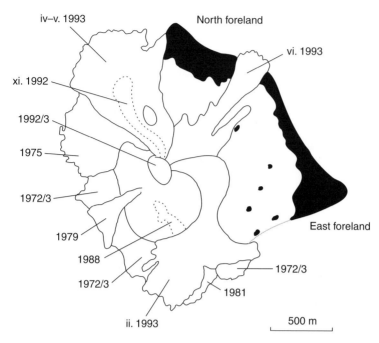

Figure 11.3 Anak Krakatau: the mosaic of substrates present by 1993, with dates of origin from volcanic activity noted; vegetation cover is shaded (cf. Figs. 11.4, 11.5). (After Thornton et al. 1992.)

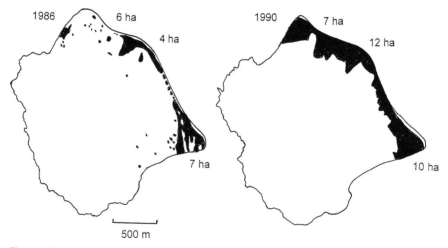

Figure 11.4 Anak Krakatau: the progress of vegetation cover (shaded) and the linking of more isolated patches, from 1986 to 1990.

eruptions in 1972–73 and involves not only plants but also birds (Zann and Darjono 1992), spiders (W. Nentwig, unpublished data), butterflies (New and Thornton 1992b), tropical fruit flies (Schmidt *et al.* 1994), and invertebrates feeding on microepiphytes on the bark and needles of *Casuarina* trees (Turner 1997) (Figs. 11.4–11.7).

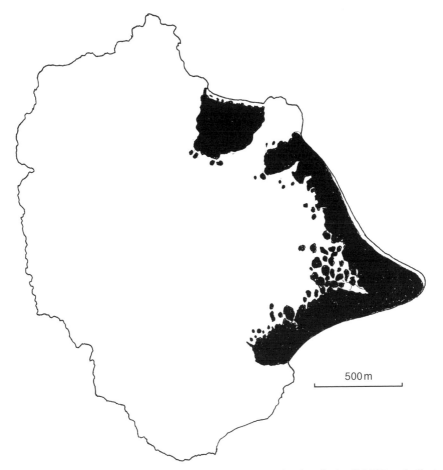

Figure 11.5 Anak Krakatau, from an aerial photograph taken in April 2005, to indicate more recent changes in vegetation cover (shaded) and consequences of recent volcanism: note extensive lava flow to north of island.

Despite the plea by Dammerman (1948) that monitoring the biota of Anak Krakatau should be accorded the 'utmost importance' and his remark that 'it is greatly hoped that this unique opportunity will not be neglected, as it was in the case of Krakatau itself', systematic surveys were not undertaken on Anak Krakatau for the quarter century following the major eruptions of 1952–53, which were presumed to extirpate earlier communities to leave simply the skeletons of taller casuarina trees projecting from a 3–5-m-deep layer of ash covering the island. The renowned Pacific botanist Raymond Fosberg visited the island in 1958, apparently the first biologist to do so, or to record his observations, since those eruptions. Although continuing volcanic activity prevented him from landing, Fosberg observed a patch of relatively large casuarinas on the north-east beach area. On a later trip (1963) Fosberg was able to land (Fosberg

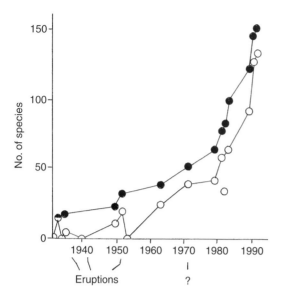

Figure 11.6 Anak Krakatau: vascular plant richness from 1930 to 1991. Filled circles, accumulative records; open circles, actual numbers of species in survey. Major known and presumed (question mark) eruptive events are shown, the first three of these extirpating the island's flora. (After Whittaker *et al.* 1992a.)

1985) and found about 25 species of plants in a grassy area at the same locality. Eighteen of those species were constituents of the island's previous early floras. Later botanical exploration in 1971 revealed the flora to have recovered substantially, to include four species of ferns and 38 spermatophytes, with four of the 15 species not found in 1963 also noted from the pre-1950s floras. The overall trend in Anak Krakatau's flora has been of increase with setbacks at intervals due to volcanic activity. Substantial doubts have been expressed (Whittaker *et al.* 1995) over whether the 1952–53 eruptions did indeed produce a 'clean slate', but both richness and coverage of vegetation increased substantially after then, so that Thornton (1996a) suggested that the island's biota in the mid-1990s represents the outcome of successful colonization since 1953, with constraints imposed by later eruptions.

Collectively, 31 of the 37 plant species presumed to occur in the floras noted in 1930–34 and 1949–51 have recolonized the island (Fig. 11.6). Partomihardjo *et al.* (1993) regarded these as a 'deterministic core' of species, most of them sea-dispersed plants of the strandline. Their survey of strandline flora during different monsoon seasons (Fig. 11.1) yielded propagules of 66 species. At that time Anak Krakatau's complement of 140 species included 58 of the 63 sea-dispersed plants occurring on the archipelago, so that almost the whole of that 'core' was already available nearby as a potential community foundation.

Successional mosaic communities, as above, give opportunities for 'nested subsets' to be present. Thus the floral successions on the north foreland and east foreland of Anak Krakatau differ in successional stage, with the former lagging the latter by about 12 years. Should one or other area be destroyed (as was the

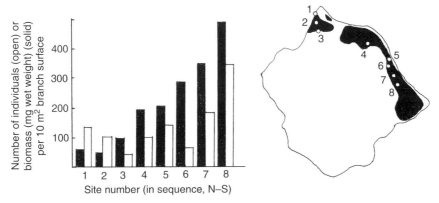

Figure 11.7 Variations in the biomass and abundance of arthropods on branches of *Casuarina* on Anak Krakatau. Upper: biomass and number of individuals at each of eight sites; lower: position of sample sites 1–8 on Anak Krakatau. (From Turner 1997.)

north-east foreland vegetation in 1993) the numbers of adjacent colonists may be biased by the condition of the remaining community or communities, should these be able to tolerate the wider environmental change wrought by disturbance. For birds, Zann and Darjono (1992) noted that the most advanced area (the eastern foreland) served as a 'bridgehead' from which birds could progressively move to the more northerly vegetated areas as these became hospitable. Although the three main vegetated regions we differentiated by name were connected almost continuously by vegetation by 1990, many birds appeared to move little. Zann and Darjono reported that, of 57 mangrove whistlers banded, only one moved the kilometre between the east foreland and north-east foreland after 8 days. Several birds were captured in 1991 in the same sites in which they had been banded in 1986.

Likewise, Turner's (1997) study of *Casuarina* invertebrates noted above implied a similar pattern. Invertebrate species richness and abundance was overall greater in the older, more southerly and eastern areas, apparently reflecting both the age of the trees and the more complex surrounding floristic environment (Fig. 11.7). Studies of other biota on the three forelands in the 1980s and early 1990s yielded other examples of such nested subsets, implicitly founded in successional state. The special case of community modules founded in figs and their mutualists is discussed seperately as a striking parallel to other tropical areas. The asynchrony of the three Anak Krakatau forelands can be attributed directly to volcanic activity, and the pattern has now been interrupted again with a lava flow from June 1993 forming a clear break which obliterated the north-east foreland vegetation and isolated the north foreland as a *kipuka* of vegetation (Thornton *et al.* 2000). Since then areas of vegetation to either side have continued to expand (Fig. 11.5).

Recent volcanic activity has also affected animals. The recent colonization of Anak Krakatau by the antlion *Myrmeleon frontalis* (see Turner 1992) was facilitated by availability of a dry sandy area in which larvae could construct pits, under the floor of a shelter. This shelter had been obliterated by September 1995, and searches failed to find the antlion; it is presumed to have been extirpated.

The presence of species after eruption has two possible interpretations. Either the species survived, or it recolonized subsequently. Some of the fig species on Anak Krakatau (such as *Ficus fulva* and *F. septica*) may be 'tolerators', and a number of vagile birds probably also fall into this category. Of the groups we studied, possibly around one-third tolerate eruptive episodes, a surprisingly high proportion that largely comprised early successional species.

The decade 1982–1992/3 showed significant changes in the complexity of Anak Krakatau's biota, with development of the most complex assemblages in the island's short history. It is hoped that scientists in the future will be able to heed Dammerman's plea for effective monitoring of their fate once the current continuing volcanic episodes subside suficiently for safe surveys to occur.

CHAPTER TWELVE

Surtsey, Island of Surtur, b. 1963

> Surtur [the Black, the fire-giant] fares from the South
> with the scourge of branches,
> The sun turns black, earth sinks in the sea.
> The hot stars down from Heaven are whirled.
> Fierce grows the steam and the life-feeding flame,
> till fire leaps high about Heaven itself.
> Icelandic poem *Voluspa*, tenth century, quoted by Sturla Fridriksson
> in his book *Surtsey: Evolution of Life on a Volcanic Island*

Emergence and development of Surtsey

The longest physical feature of this planet is almost certainly the mid-ocean ridge, a narrow deep trench running down the middle of the ocean floors. It has been likened to the seam of a gigantic tennis ball. Magma from deep inside the Earth emerges along the ridge, and the rigid tectonic plates on each side of it gradually move away from one another, as new ocean floor is created on each side of the ridge as the cooling magma becomes lava. The ridge becomes subaerial at one place in the North Atlantic, where the island of Iceland has been formed. On Iceland, continuing activity along the ridge creates new real estate, and Iceland is increasing in area.

In early November 1963, earthquakes were registered some 33 km off the southern coast of Iceland. On the morning of 14 November, 20 km from the nearest large island, Heimaey, of the volcanic Westman Islands, fishermen smelled sulphurous fumes as they were casting their lines, and saw a column of vapour rising from the sea surface not far from their boat. Then there were continuous explosions and eruptions of cinders and ash, and the vapour column soared 9 km into the sky. The fishermen had seen the first birth-pangs of a new island, whose foundations were laid down as accumulations of volcanic products on the 130-m-deep ocean floor. Next day, in view of the fishermen, the under-sea cone of ash and cinders broke the sea surface as a spectacular phreo-magmatic volcanic eruption, with sea water explosively reacting with the hot magma to send steam and cock's-tail plumes of ash far into the air at each eruptive event. A low, barren, steaming island of black ash emerged from the

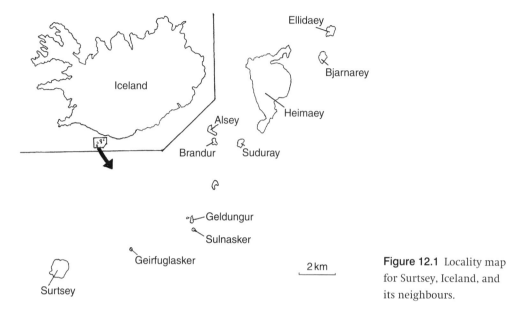

Figure 12.1 Locality map for Surtsey, Iceland, and its neighbours.

sea. The newborn island was named Surtsey, meaning island of Surtur, the black fire-giant of ancient Norse mythology. Its birth was responsible for a new term in volcanology – Surtseyan, the name for eruptions of this classic, explosive style of water–magma interaction. But the island has contributed more to science than simply the name of an eruptive type. It has shown how terrestrial ecosystems can come into being on a non-living, steaming substrate of ash and lava, many kilometres from the nearest land, in a harsh, cold–temperate climate.

The island was almost circular (Fig. 12.1), and there was a small lagoon on the north coast. Explosive activity continued as sea water constantly penetrated the crater, and marine erosion repeatedly broke down material that had taken weeks and months to accumulate. Finally, on 4 April 1964, the volcanic vent became sealed from the sea by a lining of lava, and the explosive phase of activity was replaced by an effusive phase, producing lava which, until 7 May 1965, flowed over the southern slopes into the ocean. An area of 1.5 km² of the southern side was covered in solid lava and the island's permanence thereby assured. Surtsey was here to stay.

In August 1996 activity was resumed and continued, non-stop, for 9.5 months. From a fissure and row of smaller craters to the east of the older crater, lava flowed towards the east coast. Seven months later, on 1 January 1967, lava broke through a northern cinder cone and flowed into the northern lagoon. By the end of Surtsey's eruptive period, in June 1967, the island had an area of 2.7 km². In the middle of the island a large tephra cone surrounded each of two main craters, the larger, eastern cone reaching a height of 154 m above sea level.

Sea currents carried material eroded from the shores to leeward, forming a low, mobile tephra foreland whose shape is continually changed by further erosive activity. By 1998 these erosional forces had reduced the island's area by 44%, to 1.5 km². Further south, however, effusive eruptions of the main craters produced a lava apron which covered much of the southern half of the island, consolidating its shape and ensuring the island's permanence. Most of the lava has been covered by a layer of tephra drifting from the cones to the north, but in the south-east the flows are largely free from ash cover, although airborne dust deposits have filled cracks and hollows. Surtsey's climate is said (by Icelanders!) to be 'mild and oceanic'. Mean annual temperature at Heimaey from 1961 to 1990 was 4.8 °C, and mean precipitation 1590 mm. Summer, from May until mid October, is the time of year when the island is generally frost-free.

Scientific research on Surtsey

Icelandic scientists were quick to appreciate the golden opportunities provided by their new island, and almost immediately set up a Surtsey Research Society to coordinate and facilitate research. Surtsey is the only emergent island to have been regularly and systematically monitored since its birth. Volcanological aspects were studied under the leadership of Sigurdur Thorarinsson, and later involved geomorphological, thermal and magnetic studies. Some idea of the bold, tenacious science that this island stimulated can be gained from the fact that in 1979, despite formidable logistic problems, a hole was drilled through the island from a point just inside Surtsey's first crater, then at 58 m above sea level, to a depth of 128 m below sea level, providing a 186-m-long core (82% of which was recovered) through the young volcano! But it is the biological research that has taken place on Surtsey that concerns us here.

Sturla Fridriksson's fascinating book, *Surtsey: Evolution of Life on a Volcanic Island*, published in 1975, provided a summary of the work carried out on the colonization of Surtsey in the first 12 years of its existence by him and other biologists such as C. H. Lindroth, H. Andersson, Hogni Bödvarsson, Sigurdur H. Richter, Skúli Magnússon and Hördur Kristinsson. Since then, marine studies have been set in train, and Sturla Fridriksson has continued work on the land biota together with other scientists including Borgthór Magnússon, Erling Ólafsson, L. E. and E. Henriksson, P. Gjelstrup, H. Frederiksen and Hólmfrídur Sigurdardóttir, to name but a few. The Surtsey investigations, necessarily restricted to the brief summer, and predominantly in the last week of July and early August (except 1991 when the visit was made in September), provide a fine example of foresight, attention to detail, scientific rigour and disciplined team effort.

From the beginning the decision was taken to forego Surtsey's undoubted economic potential to Iceland as a tourist attraction in favour of its value as a natural laboratory for the biological and volcanological sciences. Access to the island was permitted to scientists only. This policy has been eminently successful. The Surtsey Investigation Project is one of which Iceland can be justifiably proud; nowhere else in the world has there been a comparable long-term, regular and detailed systematic study from the very beginning. Anak Krakatau, in some ways Surtsey's tropical counterpart, was not monitored thoroughly during the early stages of its colonizations and successions, and, although part of a national park, it is visited frequently by tourists, and at the time of writing there is no permanent presence of park staff.

Surtsey and the Krakataus

Surtsey invites comparison with the Krakatau situation in a number of ways. It is about the same distance from Iceland (35 km) as was the island of Krakatau from Java and Sumatra (about 44 km). As a source area, however, Iceland stands in sharp contrast to the biologically rich islands of Java and Sumatra. It was totally denuded by glaciation during the Pleistocene, when it was covered by a 1-km-thick dome of glacial ice. Its biota is thus a young, impoverished one, itself the result of colonization over the 10 000–15 000 years since the island was freed from its thick ice envelope. There are few woods on Iceland, and wind erosion has severely restricted vegetation cover. Moreover the relatively harsh climate contrasts with the tropical conditions prevailing in Sunda Strait.

The second obvious comparison is with Anak Krakatau. The two emergent islands are remarkably similar in area, height and morphology, both having extensive lava shields that have ensured their permanence, and ash-covered areas that include highly mobile lowland promontories subjected to rapid marine erosion and deposition. Anak Krakatau's biota dates from self-devastating eruptions in 1952–53, some 22 years after its emergence, and has suffered checks on at least two occasions since then, the latest substantial eruption occurring in the 1990s. Surtsey's colonization dates from its emergence in 1963, and since 1967 there has been no self-devastating eruption. In 1965 two satellite islands emerged from the sea off Surtsey, Syrtlingur (Little Surtur) about 600 m to the east, and Jolnir (Yule) at a similar distance to the south-west (Fig. 12.2). They were both of tephra, and both disappeared within months. A further effusive phase on Surtsey lasted from August 1966 to June 1967, when less extensive lava flows were formed. Since then there has been no further volcanic eruption on the island, which by that time had grown to about the same area as that of Anak Krakatau.

Both Surtsey and Anak Krakatau have neighbouring islands carrying biotas that are subsets of those of their mainlands. In Surtsey's case the probable sources are the Westman Islands (Fig. 12.3), the closest of which, Geirfuglasker, is 5 km distant. Anak Krakatau is surrounded by its older, pre-1883, companion islands of the Krakatau group, at a distance of 2–4 km, and the island of Sebesi lies about 15 km to the north-north-east.

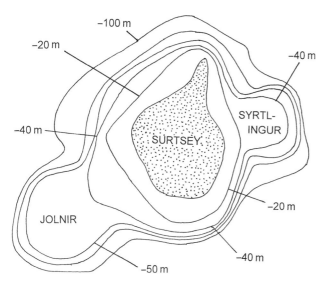

Figure 12.2 Outline of Surtsey (shaded) and subterranean contours. (After Fridrikkson 1975.)

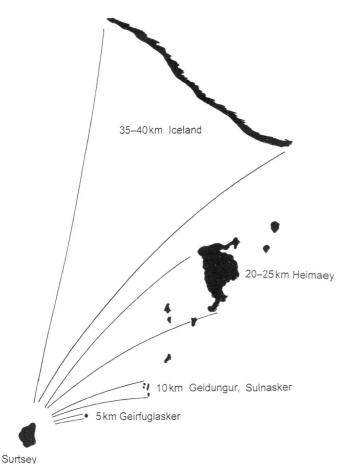

Figure 12.3 Possible sources of biota reaching Surtsey, and the distances involved.

The course of colonization

In the following treatment of the colonization of Surtsey I have drawn freely from Fridriksson's masterly 1975 book, as well as papers by him and by Borgthór and Sigurdur Magnússon on plant colonization, and publications on land arthropods by Lindroth, Andersson, Hogni Bödvarsson, Sigurdur Richter and Erling Ólafsson. A full bibliography of scientific research on Surtsey is available through the World Wide Web (see Surtsey 2006).

Colonization by plants

The whole of Surtsey was mapped as 100 m × 100 m quadrats, with each quadrat used as a recording unit. For the first few years every individual plant was meticulously marked with a numbered stake and its precise position plotted, and the invasion and spread of individual plant species was carefully monitored (Table 12.1). From 1976, when plants became too numerous to be counted, the frequency of species and extent of cover of vegetation was estimated using transects or quadrats, as well as ground and aerial photography.

Treub found that blue–green algae (cyanobacteria) were an important factor in the initial establishment of ferns on Rakata following the 1883 eruption. They were not recorded in the early stages of Anak Krakatau's colonization (the island was not carefully monitored at equivalent times since the initiation of its several successions), but they were among the first pioneers of Surtsey. Bacteria became established on Surtsey very early, even before the island-forming eruption had ended (Poonamperuma *et al.* 1967). Subsequently, what were known as 'oases of ecogenesis' (Schwabe 1971), moist places near thermal vents and craters, became foci for colonization. K. Behre and G. H. Schwabe found both free-living (*Nostoc*, *Anabaena*) and lichen-associated nitrogen-fixing cyanobacteria in 1968, and L. E. Henriksson showed that they played a role in the nitrogen economy of Surtsey's primary succession (Henriksson and Rodgers 1978). Schwabe demonstrated a fairly close association between the cyanobacteria and mosses, and both mosses and lichens have been shown to obtain nitrogen from the activities of nitrogen-fixing cyanobacteria. H. B. Frederiksen and colleagues (2000) demonstrated that the microbial community of bare sandy tephra was capable of very rapid increase in biomass whenever nutrients become available and suggested that this was an important mechanism for the retention of nutrients in bare soil for subsequent use in plant growth. During the plant succession that was stimulated by the establishment and growth of seabird colonies (see below), total microbial biomass increased significantly (to 17 times that of bare soil) and fungi, rather than bacteria, came to dominate its composition.

Two species of mosses were found on Surtsey in 1967 and identified by Bergthor Johansson; by the following year there were six, and by 1970 18 species (Fig. 12.4). In a specialist bryophyte survey in 1971, 38 species of mosses and one liverwort were found. A year later (9 years after Surtsey's emergence) a similar

Table 12.1 *Surtsey vascular flora in 1986: numbers of plants (see also Table 12.2)*

	Numbers in 1986
Species present by 1973	
Cakile arctica	0 (last recorded 1980)
Elymus arenarius	200
Honkenya peploides	(large number)
Mertensia maritima	120
Cochlearia officinalis	3
Stellaria media	0 (last recorded 1973)
Cystopteris fragilis	0 (last recorded 1984)
Angelica archangelica	0 (last recorded 1973)
Carex maritima	2
Puccinellia retroflexa	2
Tripleurospermum maritimum	1
Festuca rubra	1
More recent arrivals	
Cerastium fontanum	34 (first recorded 1975)
Equisetum arvensis	0 (recorded only 1975)
Silene vulgaris	0 (recorded only 1975)
Sagina sp.	0 (recorded only 1975)
Juncus sp.	0 (recorded only 1975)
Atriplex patula (?)	0 (recorded only 1977)
Rumex acetosella	80 (first recorded 1978)
Cardaminopsis petraea	0 (recorded 1978–1980)
Poa pratensis	1 (first record)
Sagina saginoides	150 (first record)
Armeria vulgaris	1 (first record)

Source: After Fridriksson (1987).

survey revealed an astonishing 68 species of mosses and four liverwort species. The presence of cyanobacteria in some areas of Surtsey at the beginning of a remarkable colonization by a rich moss flora recalls the film of cyanobacteria (then called blue–green algae) that accompanied and facilitated the early fern phase on Rakata, but the parallel is by no means close.

On Rakata the build-up of bryophyte species was much slower than it was on Surtsey. Although Treub found two moss species in 1886, 3 years after the eruption, no additional species were found either in the 1897 or 1906 botanical surveys. By 1922, 39 years post-eruption, there were 18 species, but the bryophyte flora did not achieve the diversity of Surtsey's 9-year-old flora until some

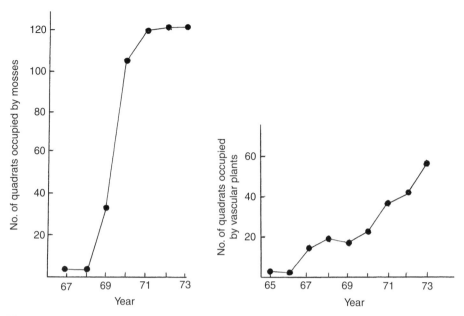

Figure 12.4 Colonization of Surtsey by mosses and vascular plants in early years from island formation (years along abscissa, from 1965 to 1973). (After Fridriksson 1975.)

time between 1933 and 1979 (50–96 years after the eruption); 32 moss and 37 liverwort species were recorded in 1983 (Jones 1986).

On Iceland, lichens are pioneers of new lava flows, particularly at high altitudes, and Surtsey has been searched carefully for lichens each summer since its emergence. It was not until the summer of 1970, however, that lichens were first found on the island. Three years later there were 12 species, and Hördur Kristinsson noticed that colonization by lichens was, surprisingly, much further advanced on the lava flows resulting from the 1967 activity than on those 2 years older (Kristinsson 1974). Aerial photographs of the island, taken in winter, had shown a thin layer of snow covering the 1967 lava but large areas of the older lava were free from snow. Kristinsson suggested that differences in heat emission, not detectable on the surface during summer but revealed by the winter photographs, may be the reason for the differential success of lichens on the 1967 and 1965 lava flows. The older lava may dry out more quickly, restricting the water available to lichen colonists. Kristinsson concluded that four lichen species, which occurred whenever there were appropriate conditions for growth, had been dispersed to the island by wind. All four form soredia, propagules which can become airborne and comprise both the green algal and the fungal components of the lichen. Three of these species harbour, in addition, the blue–green *Nostoc*, which appeared later than the green algal and the fungus components, and Kristinsson presumed that *Nostoc* was dispersed separately by wind, and then incorporated into the lichen after arrival. Free colonies of *Nostoc* were isolated from Surtsey's lava flows by Schwabe (1971), showing that the blue–green alga certainly reached the island. One lichen species was thought by Kristinsson to be bird-dispersed; many separate plants were found in an area of only 2 m², and nowhere else on the island. This particular species does not form soredia, but its erect finely branched lobes easily break off and could become attached to birds. Moreover, the lichen is coprophilous and known elsewhere to occur particularly in bird roosting sites.

As might be expected, the colonization of Surtsey by vascular plants has been relatively slow. For the first decade of the island's existence colonization rate by plants was more than one species per year, but the second 10 years seemed to be a period of consolidation. With the flora at about 24 species, few new species arrived and colonization rate dropped sharply. For the first 20 years most of the colonizing plants were species of coastal, sandy habitats, probably dispersed by sea or by birds. In the third decade, however, a significant change took place with the establishment and expansion of the seagull colony. Plant colonization rate rose to about three species per year, but even by 1992, 29 years after Surtsey's emergence, the island carried only 30 plant species, or still only about one successful colonist per year. Most of the vascular plants that colonized Surtsey in the first three decades of its existence were sea- or bird-dispersed coastal or bare open ground species.

In the last decade a dense patch of nitrophilous plant species has developed in the gull colony on the southern lava field, what Sturla Fridriksson describes as 'much like any assortment of a barnyard flora'. By 1999, 45 species of vascular plants were growing on the island, making a total of 55 recorded since its inception, and in the oldest part of the seabird colony the vegetational cover was complete (B. Magnússon, unpublished data). About 85% of the 55 known species are permanent members of the Surtsey flora, and, of these, Fridriksson (2000) estimated that 75% may have been brought by birds, 11% by sea and 14% by air. About 80% of species occur on the other Westman Islands.

Of the 55 species known to have grown on Surtsey between 1965 and October 1999, 35 persisted to the summer of 1995, 47 to the summer of 1998 and 44 to 1999 (Magnússon *et al.* 1996, Fridriksson 2000, Magnússon and Magnússon 2000). Sturla Fridriksson believed that in the early years extinctions following initial establishment were largely the result of the unsatisfactory nature of the loose dry sand, hard lava and gradually solidifying tuff as substrates, and since the establishment of the gull colony species were lost because of the rich fertile soil and competition with later, more aggressive colonists. Most of the losses have been plants that were established in the gull colony. In spite of this turnover, Surtsey now has more species than any of the other Westman Islands, but although several Surtsey species have not been found on those islands Surtsey's flora is comparable to that found elsewhere in the Westman Islands region.

Although a number of viable seeds of vascular plants were collected on Surtsey's shores during the first summer (1964) of the island's existence, it was not until the second summer that Fridriksson found higher plants growing on the island. In June 1965 he found seedlings of an annual, the sea rocket, *Cakile arctica*, in a mixture of tephra and decaying thalli of the seaweed *Ascophyllum nodosum* on the sandy beach in the north of the island. The nearest colony of sea rockets is on Heimaey, 20 km from Surtsey. A few weeks after their discovery the Surtsey seedlings were killed by ash falling as a result of the Syrtlingur eruption. The sea rocket was found growing again in 1966, 1967, 1969, 1972 and 1973. Another species of *Cakile*, *C. maritima*, was found on the emergent island Nea Kameni in the Mediterranean (see above), and yet another, *C. lanceolatum*, was a pioneer of the sandy marine cay Cayo Ahogado, off Puerto Rico, studied by Heatwole and Levins (see below). Sea-dispersed species of this genus appear to be successful early colonisers of new islands in warm temperate and tropical as well as sub-arctic seas.

In the summer of 1966 a second colonization attempt occurred. Four seedlings of lyme grass, *Elymus arenarius*, were found growing, again on the sandy north shore, at the high-tide line. They were destroyed by waves after a few weeks. In 1967 a third, more successful colonization took place by no fewer than 26 individuals of the sea sandwort, *Honkenya peploides*, a common perennial on Iceland's southern shore. One specimen of the lungwort, *Mertensia maritima*, the fourth vascular plant species, was discovered. Lyme grass and sea rocket were also present, and 15 sea rocket plants flowered, six of them setting about 300 pods collectively with about 600 seeds.

In 1968, surprisingly in view of the previous year's success, no sea rockets were found, but the sea sandwort had increased in numbers and extent. In 1969 a fifth higher plant species, scurvy grass, *Cochlearia officinalis*, had arrived, the location (bird droppings near rainwater barrels) showing that they had been carried by birds.

A sixth plant arrival, the common chickweed, *Stellaria media*, was found in 1970. Four individuals were found growing in a sand-filled lava crevice surrounded by pieces of egg shells, feathers and bird droppings. They set 12 fruits, some of the seeds ripening and dispersing in late summer. In 1971 three small individuals of the bladder fern, *Cystopteris fragilis*, were found growing down from the roof of a lava cave. This is Iceland's commonest fern, growing on rocks, cliffs and lava flows.

In 1971 an important point in plant colonization was reached. Four years after it was first discovered on the island and 2 years after it was first found flowering, five of the 52 plants of the sea sandwort flowered and one produced seeds. No longer was the future population growth of this species dependent on an annual influx of seeds arriving accidentally from outside. By 1976, 73 sea sandworts flowered, one study plant producing about a thousand pods. Soon, hundreds and then thousands of new plants grew annually, and the species is now widely distributed over the island's sand-covered lava and is Surtsey's most common vascular plant species.

In 1985 a change occurred in the colonization of Surtsey which was to prove of the utmost importance. Herring gulls (*Larus argentatus*) and lesser black-backed gulls (*L. fuscus*) began nesting on the southern lava apron and established a mixed colony there. The colony grew year by year, and since its establishment the number of new plant arrivals to the island has increased sharply. Whereas before the colony was established nitrogen for plant growth must have been derived from sea spray, the atmosphere, nitrogen-fixing microorganisms, beached organic flotsam and occasional bird droppings, in the area of the gull colony there was now a steady supply of nutrients, including both nitrogen and phosphorus, through fish scraps, regurgitated pellets, dead chicks (and adults) and bird droppings. In this area, soil fertility and microbial activity and biomass increased dramatically, and fungi became an important component of the soil community.

Borgthór and Sigurdur Magnússon (2000) followed soil development and the succession of vegetation in 25 permanent plots, 10 in the gull colony and 15 outside it, selected for different substrate conditions and seabird influence, from 1990 to 1998. During this period there was a steady increase in plant cover and species richness on Surtsey, no fewer than 26 plant species colonizing, a much higher rate of colonization than previously (Table 12.2). The increases were confined to plots in the gull colony, except for one plot near a small crater where fulmars (*Fulmarus glacialis*) were nesting in increasing numbers. *Honkenya* increased its cover in sandy parts of the gull area to 30–50%.

Soil outside the gull colony had a high pH (7.5) and low total carbon (<0.1%) whereas within the colony pH was lower (6.4) and total carbon ten times higher (1.1%), although much lower than the carbon content of grassland soils in Iceland (5–15%).

In the nutrient-poor habitats outside the colony, sandy areas had already been colonized by a few coastal clonal perennials, and the Magnússons found that changes in cover and species richness were minor. *Honkenya peploides* was the dominant plant, together with *Elymus arenarius* and *Mertensia maritima* fully colonizing sandy areas.

Plant cover and species richness both increased two- to threefold within the colony over the 8-year study period. But improved nutrient status of the soil cannot explain the increased colonization of new plant species within the gull colony, for such soil improvement also occurred at the more dispersed inland nest sites of the greater black-backed gull (*Larus marinus*) (which also occurs within the gull colony) and the fulmar. These inland nest sites have never been foci for the colonization of new plant species. The difference is thought to lie in the different feeding habits of the two groups of seabirds. Lesser black–backed and herring gulls, the two dominant species in the gull colony, in Iceland

Table 12.2 *Early development of the vascular flora of Surtsey: numbers present each year (1965–1973)*

Species	Year								
	1965	1966	1967	1968	1969	1970	1971	1972	1973
Cakile arctica	23	1	22		2			1	33
Elymus arenarius		4	4	6	5	4	3		66
Honkenya peploides			24	103	52	63	52	71	548
Mertensia maritima			1	4				15	25
Cochlearia officinalis					4	30	21	98	586
Stellaria media						4	2	2	1
Cystopteris fragilis							1	4	3
Angelica archangelica								2	2
Carex maritima								1	1
Puccinellia retroflexa								2	1
Tripleurospermum maritimum								1	5
Festuca rubra									1
Unidentified plants				1			4	2	1
Total plants	23	5	51	114	63	101	83	199	1273

Source: From Fridriksson (1975), some nomenclature updated after Fridriksson (1987).

feed on insects and earthworms in grasslands and hay fields, and may also take up plant seeds, either directly or within the bodies of earthworms. In worm casts and the gut of earthworms seeds have been found of several of the same species (including *Sagina procumbens, Cerastium fontanum, Stellaria media, Poa annua* and *Capsella bursa-pastoris*) and genera (*Agrostis, Taraxacum, Rumex*) that have been found growing in the Surtsey gull colony. Gulls have been shown to be dispersers of viable plant seeds (including those of *Polygonum aviculare, P. annua, S. media* and *Cerastium, Festuca* and *Rumex* species, all found in the gull colony) over at least 15–20 km in the Pembrokeshire islands in the UK (Gilham 1970). Both lesser black-backed and herring gulls use plant material for nesting, and with little available on Surtsey may have had to visit other islands for nest material. After 1993 the gull colony expanded considerably and the herring gull became the most common breeding bird. The young are fed on vegetable matter as well as fish, and plants are brought in in this way.

In contrast, the greater black-backed gull is more a shore and pelagic feeder on fish, carcasses and flotsam, and the fulmar never feeds on land. The former includes plant material in the nest but the fulmar does not.

Most plant colonists of the gull area have been perennial and annual herbs with abundant seed production, well adapted to disturbed, fertile habitats and found in seabird colonies on other Westman Islands. Colonization and spread of the pearlwort, *S. procumbens*, has been rapid and it is the main pioneer of laval areas both within and outside the gull colony. Within the colony, the most abundant species, apart from *Honkenya*, are *P. annua* and *Puccinellia distans*. Other species in the colony, where the average cover is about 30%, include *C. fontanum*, *S. media*, *Poa pratensis* and *Cochlearia officinalis*. It was only after the formation of the gull colony that annuals, such as *P. annua* and *S. media*, became firmly established on Surtsey. One of the latest plants to be found, in July 2001, is eyebright, *Euphrasia frigida*, which grows in dry meadowland on some of the outer Westman Islands, Heimaey and the south coast of Iceland. Vegetative parts of a species of *Euphrasia*, very probably *E. frigida*, were found, drifted ashore, on Surtsey in 1964, a year after its birth, but Sturla Fridriksson believes that the successful immigrants found thriving in the area of the herring gull breeding colony were almost certainly brought as seeds by the gulls.

The Magnússons have found that in recent years changes in cover and composition in the plots in the centre of the colony have slowed, and in some of the lava plots the *S. procumbens* cover has declined. They believe that future development of vegetation in the colony may lead to a community similar to that in the species-poor grassland and forb communities characteristic of the outer Westman Islands. All but one of the component species of these communities have already colonized the gull colony on Surtsey, but in the Westman Islands grassland communities *Festuca richardsonii* is the dominant species, whereas in the Surtsey gull colony this species is less common than *P. annua* or *P. distans*. However, where it does occur it forms dense patches that are expanding rapidly.

It was not until 1993 that the first woody plant, the crowberry, *Empetrum nigrum*, was found on Surtsey, but three willow species colonized shortly afterwards, *Salix herbacea* in 1995, *S. phylicifolia* in 1998 and *S. lanata* in 1999 (Sturla Fridriksson 2000 and personal communication). These species do not grow on the other small islands of the Westman Group, and Fridriksson believes that they must have arrived as seeds carried from the mainland of Iceland, and were able to become established on Surtsey, where there is little competition.

Infertile areas of tephra sand, or lava and tephra sand, away from bird colonies, have been colonized by 13 species of perennial clonal herbs with extensive roots, such as *Honkenya peploides*, *Elymus arenarius* and *Mertensia maritima*. Of these, *H. peploides* is the most widely dispersed, reaching a cover of 20% in some sandy areas, and is the most successful colonist. Plant cover away from bird colonies averages about 4%. Lava fissures and hollows are the habitat of the fern *Cystopteris fragilis*, and in 1995 *Juncus alpinus*, *Salix herbacea* and *Galium normanii* were recorded in this habitat, although it is not yet known if these three species have become established.

The surveys of Fridriksson, the Magnússons, and their colleagues have been so regular and frequent (every summer), and their work so detailed, that arrivals failing to become established could be recorded even when they persisted for only one year. Four species failed to colonize after arriving once, but nine have arrived more than once, and one has reached Surtsey as many as five times without becoming established. In all, 13% of the species that have been found on Surtsey failed to become permanent residents, a surprisingly small proportion when one considers the harsh climate, and large areas of infertile substrates such as lava, solidifying tuff or loose dry sand which potential colonists must encounter.

Colonization by animals
Six months after Surtsey's emergence Fridriksson collected a fly on the shore, and in the next summer more flies and midges, two species of moth and a mite were collected. The mite appeared to be feeding on dead midges, which in turn had been living on a bird carcass, and a mould was found on one of the mites. Thus in this short time a food chain had developed from the carcass through the midge and mite to the mould.

Regular trapping of arthropods was instituted in 1966, using sticky traps, and the systematic collection of terrestrial arthropods began. In 1966, 22 species of insects and four of arachnids were recorded. The number of species increased annually, until by 1970 158 species (over 5000 individuals) were recorded, almost one-fifth of the Icelandic fauna. By 2002, about 300 species of invertebrates had been recorded from Surtsey.

The invertebrate fauna of the Westman Islands was surveyed in 1968. It appeared that Heimaey had been the source of most individuals arriving on Surtsey, but it was shown that some species must have travelled from Iceland, and others, active fliers like noctuid moths and some butterflies, from Europe.

Very few of these early arrivals were able to become established on Surtsey. In 1975, Fridriksson considered that only three terrestrial invertebrates had successfully colonized the island, all feeding and breeding on washed-up carcasses: a fly, *Heleomyza borealis* (first found in 1965), a collembolan, *Archisotoma besselsi* (found in 1967 and capable of being dispersed on the ocean surface), and a chironomid midge, *Cricotopus variabilis*, common on Heimaey.

In 1968, tephra samples were examined for protozoans, and flagellates were found, but at much lower density than in soil of an Icelandic grassland community. In 1970, vegetation samples were cultured and rhizopod and ciliate protozoa, as well as two species of rotifers, were discovered, evidently feeding on bacteria and cyanobacteria.

The Swedish nematologist Bjorn Sohlenius identified three nematode species in 1971. One of them, *Acrobeloides nanus*, is a cosmopolitan, parthenogenetic, unselective bacterial feeder, able to survive desiccation. The others, species of *Monohystera*, are also bacterial feeders. In 1972 *A. nanus* was again collected, with a *Monohystera* species and *Plectus rhizophilus*, a cosmopolitan species known to be extremely resistant to changes in temperature and water availability. *Plectus rhizophilus* was collected again in 1973. The first soil nematodes were found in 1972, close to the shore in soil from under a piece of wood covered in fungus. They were of the genus *Ditylenchus* (Sohlenius 1974).

On the Krakataus, four desiccation-resistant nematodes were found in moss on Rakata in 1921 by Dammerman, and included a species of *Plectus*. R. Winoto Suatmadji and his colleagues identified at least 77 nematode genera from the archipelago in 1985, including the genera found on Surtsey. Among the 19 genera on the young emergent island Anak Krakatau, the bacterial-feeding Cephalobidae were well represented (by four genera) and included *Acrobeloides*, as were the Tylenchida (also four genera) including the fungivorous *Ditylenchus*. Both these genera were found on all islands. *Monohystera* and *Plectus* were not found on Anak Krakatau (Winoto Suatmadji *et al.* 1988).

In 1978, one braconid and four species of ichneumonid Hymenoptera were found on Surtsey. Three years later, Erling Ólafsson (1982) found 40 species of arthropods other than Collembola on Surtsey. These comprised the following: an aphid species (on the *Honkenya* plants); a *Plutella* moth; a carabid and two staphylinid beetle species; 28 species of Diptera of 15 families, including among others chironomids, a ceratopogonid, a phorid, coelopids, heleomyzids, ephydrids, a sphaerocerid, a drosophilid, carnids, scatophagids, muscids, anthomyiids and a calliphorid; five species of linyphiid spiders; and two mite species. He also found an enchytraeid oligochaete worm.

Mites, enchytraeid worms, nematodes and collembolans (springtails) form the soil 'mesofauna' and, together with earthworms, are important decomposers, breaking down plant material and increasing its availability to microorganisms. The microbial activity releases nutrients into the soil where they become available to plants. Again, the gull colony has been important, in facilitating the development and dispersal of the soil fauna, which is still increasing and changing.

Mites (of the Gamasida) were first recorded on Surtsey in 1965 when one was found on a midge, and in the following year oribatid mites were found on driftwood. Most subsequent finds of mites were in association with the scientists' small hut which had been erected on the island. Lindroth *et al.* (1973) and Bödvarsson (1982) recorded 16 species of mites (only one an oribatid), equal to the number of collembolans. The mites were thought to have arrived through the agency of flies, birds, driftwood, strong winds and humans, but the presence of a permanent soil fauna on the island was not confirmed until 1995. In that year Gjelstrup (2000) and Hólmfríður Sigurdardóttir collected data on the soil community from five 3×3 m squares, each representing different successional communities. They found that turnover had been great.

Forty species were present, but only two of the 16 species previously known were found, one being common and widespread. The remaining 38 species were part of a new soil fauna, established since 1978. Two species (of oribatids) have

near relatives in North America and may have colonized from the west; the rest probably originated in Iceland or western Europe.

The highest densities, of 80 000–240 000/m^2 and 68 000–97 000/m^2 mites and collembolans, respectively, were found in the gull colony, in soils dominated by the grasses *Puccinellia retroflexa*, *Poa pratensis* and *Cochlearia officinalis*. In a plot within a pioneer dune community of *Honkenya peploides* and *Elymus arenarius*, some 190 000 mites and 10 000 springtails per square metre were found. Although overwhelmingly dominated by a new species of *Hermannia*, oribatid mites, which are mostly feeders on fungi or bacteria, were found to be surprisingly diverse in vegetated areas; they were represented by 22 species, all new to Surtsey. Fifteen species were found in the gull colony and two were even present in the control area devoid of vegetation. There were nine species of Actinedida; one, which may feed on plants or fungi, was easily the most numerous mite on Surtsey, with more than 11 000 specimens collected. Others of this group are plant and algal feeders or are probably predaceous on other soil organisms, and all but one shore-inhabitant lived in soil. Four species were Acaridida, a group characterized by a highly mobile second-instar larva equipped with suckers for attachment to animals, including insects, which act as dispersers. These acaridids were particularly numerous in the area of the gull colony. There were six gamasid mites, which are predaceous, sometimes attaching to insects, the two commonest being especially prevalent in the gull colony and known from Heimaey. Gjelstrup (2000) believed that as more stable plant communities develop, the mite fauna will change and become reduced.

In 1976 Collembola species were found for the first time, four species occurring in samples of moss from the southern lava fields. All are very common inhabitants of moss in America and of soil in Iceland and Heimaey, and three occur on one or more of the nearby small Westman Islands. In 1978, when, as in previous years, 30 samples of moss were taken for extraction in a Berlese apparatus (a simple device whereby a combination of heat and light gradually forces small invertebrates out of the vegetation or litter sample into a vial of alcohol), no collembolans were present, and Bödvarsson concluded that the moss fauna had been eradicated, either by the extreme drought of the 1978 summer, or by being covered with wind-blown tephra. By then, seven species had been found on the flat, sandy, northern foreland of Surtsey at places away from the shore.

Hogni Bödvarsson noted in 1981 that of the 16 collembolan species known from Surtsey, by 1978 at least four, and possibly nine, had become permanent inhabitants. Twelve species were found on the shore and three were definitely halobionts (species of salt environments). Most of the others were ubiquitous forms, and four of them were found in tufts of grass that had been washed ashore. The occurrence of one species, *Vertagopus arborea*, on Surtsey was something of a puzzle. It was found on the shore in 1970 and further inland in 1972, but was not found at all in the surveys of 1974, 1976, 1978 or 1981. This species is not known from Iceland itself, but occurs on the small island of Bjarnarey some 3 km east of Heimaey; it also occurs in Europe. Intriguingly, in Europe the species is usually found on tree trunks, but trees are uncommon on Iceland and absent from both Bjarnarey and Surtsey.

We have already noted that the establishment and growth of the gull colony probably had a significant effect on the soil fauna of Surtsey. In 1995, some 10 years after the formation of the colony, Hólmfrídur Sigurdardóttir (2000) made a detailed study of this effect by comparing the distribution of springtails inside and outside the colony. No collembolans were present in the unvegetated reference plot but in sparsely vegetated areas outside the gull colony their density was 806–935/m^2, and the dominant species was *Mesophorura macrochaeta*. In contrast, within the colony the density was 17 680–74 724/m^2, comparable to that in grasslands in southern Iceland, and a different species, *Hypogastrura purpurescens*, dominated. The presence of a gull colony was clearly having a major effect on the numbers and composition of the collembolan soil fauna. Four springtail species were found new to the island.

Enchytraeid worms were first found on Surtsey in 1972, and Hólmfrídur Sigurdardóttir found them in all her study plots except the unvegetated reference plot. She also found the first earthworms on Surtsey in 1993, in samples taken from the gull colony. They were juveniles of *Lumbricus castaneus*, normally an inhabitant of litter and highly organic soils. As noted above, the two main gull species in the colony feed in Iceland on insects and earthworms, but earthworms do not survive passage through the gut of a gull. Sigurdardóttir (2000) suggested that earthworm cocoons could have been brought in soil on the birds' feet or feathers. *Lumbricus castaneus* is one of 11 earthworm species found in Iceland and occurs on Heimaey. In spite of a thorough search of the area of the 1993 discovery 2 years later, no earthworms could be found, and they have not apparently been found subsequently. In 1998, slugs were found on Surtsey for the first time, in a dense patch of grass in the gull colony, and were present also the following year. They are similar to a species common in grassland and gardens in southern Iceland. Clearly the increased nutrient status of soil in the colony affects the soil fauna, as well as the microorganisms and plants, both qualitatively and quantitatively.

The land vertebrates of Surtsey are all birds, and most are sea birds. Gulls roosted on the new island's shores when it was only a few weeks old, and in the following year a flock of kittiwakes, *Rissa tridactyla*, frequently rested on the southern lava cliffs or on steep tephra slopes in the north. In addition, greater black-backed gulls, herring gulls, glaucous gulls (*Larus hyperboreus*) and Arctic terns (*Sterna paradisaea*) were present on the flatter ground during Surtsey's first summer.

Seabirds began nesting on the island in 1970, when fulmar and black guille-mot, *Cepphus grylle*, each built a nest on the southern lava cliffs. In the next year there were 11 fulmar and six guillemot nests, and since then both species have bred regularly on Surtsey. A pair of greater black-backed gulls nested in 1974, and the kittiwake in 1975. By 1981 the herring gull was nesting, five pairs of great black-backed gulls had produced young on the island and a pair of herring–glaucous gull hybrids were breeding, their nest being made largely of moss. Lesser black-backed gull and herring gull began nesting on the southern lava apron in 1986, the beginning of a gull colony that was destined to change the rate and nature of the island's colonization, having important effects on the development of the island's flora through its soil fauna and microbial flora (see above). By 1993, the glaucous gull was also breeding, making seven species of seabirds known to nest on Surtsey each year by 1999, when the herring gull had become the commonest bird on the island, with some 300 pairs.

The role of these birds in plant colonization depends on their colony size, nest site and nest type (Magnússon and Magnússon 2000). Fulmar and black guillemot, of which there were 120 and 15 pairs, respectively, in 1990, build no real nest. Fulmars began nesting largely on the sea cliffs, but for the last 15 years small groups of 5–15 pairs have nested in five places on the cliffs of the old craters. Vegetation cover has increased at most of the crater cliff sites but has included no new plant colonists. Although large numbers of kittiwakes roost in the northern parts of Surtsey where they enrich the soil by their droppings, both kittiwake and guillemot nest on the island's sea cliffs. The kittiwake does incorporate vegetation into its nest, but vegetation has not become established at the nest sites of either species.

As noted above, the gull colony on the southern lava apron was initiated in 1986, by the first nests of lesser black-backed gulls. The colony was estimated in 1999 to comprise about 300 pairs, and included three other species, herring, greater black-backed and glaucous gulls, mostly nesting within about 10 m of one another over an area of about 7 ha. The nests are made mostly of plant material. Moss was used in the early years by the lesser black-backed and herring gulls, but grasses and *Honkenya* are now used. This colony has had an important effect on plant colonization on Surtsey. Its establishment and growth have led to an increase in plant cover in the area and a sharp rise in the colonization rate of plant species, most new finds being within the area of the colony (see below).

The puffin, *Fratercula arctica*, inhabits the small, 50-m-high cliff stacks of the outer Westman Islands in great numbers, digging deep nesting burrows in the deep grass-covered soil on their summit slopes. Grassland honeycombed by the network of puffin burrows takes on a special character. Its high organic content, intense aeration and high water content makes it extremely fertile, favouring the grass *Festuca rubra*, which becomes dominant. Puffins need soil of at least 1 m depth for their burrows, and a slope facing the sea from which they can push off in a downward direction to become airborne. In 2001,

Sturla Fridriksson and Borgthór Magnússon noticed that a pair on Surtsey appeared to be beginning to investigate newly formed turf as a possible nesting ground, and this may be the next seabird to colonize the island. They were reported breeding for the first time in 2004.

It was not until 1995 that the first land vertebrate bred on the island. A subspecies of the snow bunting (*Plectrophenax nivalis nivalis*) migrates from the British Isles to Greenland, via Iceland, and individuals have visited Surtsey in some years since 1964, bringing seeds to the island (see below). The bunting first bred on Surtsey in 1995, the eighth bird, and the first non-marine bird to do so; it was still breeding there in 2001, when a nest was found for the first time. In 2001 another non-marine bird was breeding: the wagtail, *Motacilla alba*, a migrant that had been noted as frequent, but not breeding, in 1967, was found with young. An insectivore, it may well find sufficient food on Surtsey year round, and become established.

Dispersal to Surtsey
Dispersal by sea
Sturla Fridriksson tested seeds of Icelandic plants, three coastal and three alpine species (five of them in fact occurring on the south coast of Iceland), for viability after varying periods of immersion in sea water. After immersion for 16 weeks all species showed more than 50% germination and one, an alpine species, a 90% success rate. After 32 weeks (8 months) of immersion, one species had completely lost its viability, but seeds of the others germinated with success rates varying from 38% to 74%. Two of the coastal species tested were present on Surtsey; they had been found there in beach drift only 6 months after the island's emergence, one of them (*Cakile arctica*) being the first plant species to grow on Surtsey, in 1965. The other, *Mertensia maritima*, became established in 1972. These tests showed that dispersal by sea from Iceland to Surtsey was certainly possible.

Of the seeds and other diaspores of some 38 species of vascular plants found on Surtsey's shores in the period 1964–72, 21 (55%) grow on the outer Westman Islands that are nearest to Surtsey, 31 (82%) occur on Heimaey, and all grow in Iceland. The species growing on the outer islands, close to Surtsey, were the most abundant in the drift material. Only 6 months after the island's emergence, leaves and stems of lyme grass and the rush *Juncus balticus* were found in beach drift, and seeds of lyme grass, of *Angelica archangelica*, and of the sea rocket, *C. edentula*, were collected. Whole green plants of *Matricaria maritima* and *Sedum rosea* were found on the shore, and a plant of the former was potted and kept alive for some time. Stolons of lyme grass are capable of developing in Surtsey's sandy soil.

In May 1969, a number of mermaid's purses, the empty chitinous egg capsules of the ray *Raja batis*, drifted ashore on Surtsey, and were found to have seeds attached to their rough outer surfaces. The seeds proved to be of four grasses common in Iceland and also found on Heimaey, only one of which, lyme grass, grows on the smaller Westman Islands close to Surtsey. The three other species were not represented among propagules found washed up on Surtsey. Fridriksson presumed that the seeds had come into contact with the egg capsules somewhere at least 20 km away (the distance to Heimaey) and floated with the capsules to Surtsey. The prevailing current around Surtsey is to the west, and at the surface may reach a speed of more than 7 km/day, so drift from Heimaey may reach Surtsey in two and a half days if there were no wind and the current set directly for Surtsey. These ideal conditions seldom occur, however, and in the summer of 1967 a large-scale trial was made in an attempt to demonstrate drift from Heimaey to Surtsey.

Several million buoyant grains of yellow plastic, 2–4 mm in diameter, were released into the sea at Heimaey. Within a week, in spite of shifting winds, a few hundred reached Surtsey, at an average drift speed of about 2.5 km/day. Fridriksson also calculated the relative chances of dispersal to Surtsey of floating diaspores from the seven islands of the Westman group closest to Surtsey, based on the proportion of the circular dispersing front that is taken up by the Surtsey sector, and the circumference of the putative source islands. Heimaey, with six times the vascular plant species numbers of the next most floristically rich island, also had about ten times the calculated likelihood of propagules dispersing to Surtsey, in spite of the fact that it is the most distant of the islands considered. This conclusion was supported by the known distribution of the eight species of Surtsey collembolans (see below).

Eleven individuals of a mite species were found in 1966, alive on a gatepost that had probably drifted from Heimaey. A flightless weevil found on Surtsey is abundant on Heimaey, but tests showed that it cannot survive immersion in the sea for more than 24 hr; it, too, probably rafted to Surtsey.

In the early 1970s Lindroth's group stressed the importance of water-borne dispersal in the island's early colonization by invertebrates, and this was supported by a number of observations (Lindroth *et al.* 1973). In 1971 six puparia, of three species of shore flies, were found washed up in drift on the northern beach. From one, that of a coelopid (a seaweed fly), a parasitic chalcid wasp emerged. The wasp was known from the Westman Islands, but not from Iceland. Ólafsson (1978) believed it was likely that the parasite had survived the ocean crossing within the fly's puparium. In the 1972 season a small grass tuft was found washed ashore on Surtsey; it harboured two live collembolans and some 30 mites. Two years later a larger tussock (90 cm long and 10–20 cm in diameter) was washed up. Half of it was subjected to detailed inspection, including extraction in a Berlese apparatus; no fewer than 663 individual land invertebrates inhabited this half-tussock, and included 398 mites and 214 individuals of one springtail species. Four other springtail species, a proturan, a chironomid larva

Table 12.3 *Dispersal spectra of vascular plant assemblages reaching Surtsey, Rakata and Anak Krakatau at comparable times (in years, brackets) since the tabula rasa, and reaching Taal volcano after 6 years*

	Dispersal mode (%)				
	Birds/bats	Sea	Wind	Humans	Total species
Surtsey 1990 (+27)	64	27	9	0	30
Surtsey 1998 (+35)	75	11	14	0	54
Rakata 1908 (+25)	23	55	22	0	110
Anak Krakatau 1979 (+27)	15	48	33	4	44
Taal 1917 (+6)	45	8	36	6	303

Source: Data from Fridriksson (1992, 2000), Thornton (1992a) and Brown *et al.* (1917).

and an enchytraeid worm were also present. Pollen analysis of the tussock's soil revealed that the pollen was 93% grass pollen and 3–4% pollen of the families Compositae and Caryophyllaceae, proportions that are characteristic of the flora of bird cliffs and coasts on the Westman Islands. The grit in the tussock proved to be a type of tuff found in the Westman Islands, particularly on northern Heimaey. Two of the halobiotic species of springtails have been seen frequently in numbers floating on the surface along the shore, and experiments reported by Lindroth and his group (1973) showed that these species, *Archisotoma besselsi* and *Isotoma maritima*, may easily be carried to Surtsey by surface currents from known sources on other islands of the Westman group and Iceland. Both halobionts and non-halobionts were found in grass tussocks washed ashore on Surtsey, showing that rafting is also possible. The non-halobiont soil collembolans were found living in moss on the inland lava fields and this was thought to indicate that these species arrived by aerial dispersal.

Sturla Fridriksson believed that three vascular plant species probably arrived by sea from the small islet Geirfuglasker, 5 km distant, two from islets 10 km distant, and seven, including the two important pioneers, lyme grass and sea sandwort (Table 12.3), from Heimaey.

Dispersal by air
Fridriksson reasoned that the relative probabilities of air dispersal to Surtsey from the Westman Islands would be related to the areas of the various putative source islands rather than their circumferences. This would favour Heimaey as the source to an even greater extent than it did for dispersal by sea. Species of ferns, mosses and lichens were probably carried by winds from Heimaey or the Icelandic mainland, and the recent colonization by three willow species was

almost certainly the result of seeds being carried by air to Surtsey from the mainland of Iceland (see above).

Flying animals (birds and insects) and many spiders are dispersed by air. In 1966, a linyphiid spider arrived, evidently by 'ballooning' on a long thread secreted by the immature spider. All the insects found on Surtsey by 1970 were winged and included an individual of the noctuid moth *Autographa gamma*, which migrates from Europe and is not native to Iceland. During strong northerly winds the number of insects caught greatly increased. Some 77 species of flies are thought to have arrived by flight, assisted by wind, and by 1999 16 species of birds had been recorded on Surtsey, eight of which were breeding there.

Dispersal of plants by birds

Only two non-migrant, non-marine bird species have been recorded from Surtsey, the raven, *Corvus corax*, first seen in 1965 and still (2000) present each summer without breeding, and the snow bunting, first recorded in 1964. A pair of ravens has visited the island regularly since 1965, and even attempted to build a nest in one of the two major crater areas. In 1968 one plant of the thrift, or sea pink, *Armeria maritima*, was found at this locality, and was flowering. The ravens had brought nesting materials to the site, and Fridriksson believed that seeds of the thrift were brought to the island in this way. By 1990, there were two plants at the site, both in flower. Being an omnivore and feeding on carrion, the raven can exploit resources such as washed-up carcasses that derive from outside the island itself, and a pair was still present in 1999, although evidently the resources are as yet insufficient to support a continuously resident pair that breeds on the island. Utilization of extrinsic energy sources by island animals has also been noted on Anak Krakatau (see below), and on Surtsey. Of course seabirds, the only other vertebrates on the island, also derive their sustenance from the surrounding ocean.

Seabirds were the first birds to land on Surtsey, feeding on carcasses washed ashore and sometimes bringing their catch on to land to feed. It is probably no coincidence that the number of vascular plant species rose sharply between 1971 and 1972 (Fig. 12.4) because of nutrient enrichment following the establishment of the first seabird colonies (fulmar and black guillemot) in 1970, and again in the decade following 1986, when the gull colony was formed.

By 1999 the gull colony was flourishing, and in the oldest part of the colony there was complete vegetative cover (B. Magnússon, personal communication). It is in this rapidly growing community that most of the vascular plant colonists since 1990 have become established, and at a rate (26 species in 8 years) very much higher than at any other time in the island's short history. Fridriksson's group examined pellets regurgitated by gulls at the breeding site, and found that they contained about 3% by weight sand, 8% plant material, mostly leaves and hulls of lyme grass plants, 10% feathers and 80% fish. On a flat surface in the colony a few procumbent pearlworts, *Sagina procumbens*, were joined by grasses, and soon there was a dense association of nitrophilous species. This plant community's rapid development since 1985 is the result of the activity of lesser black-backed gulls and herring gulls carrying seeds to their nest sites and increasing soil fertility with their guano and food waste.

Fridriksson (1992) estimated that 64% of the vascular flora had arrived through the agency of birds, 27% by sea and 9% by air. Of the 30 species of plants present in 1992, 14 were restricted to the small areas of the island used by birds for breeding or roosting, ten of them to seabird colonies. As a result of the part played in plant dispersal by the seabird colony, Fridriksson (2000) revised these estimates to 75% (birds), 11% (sea) and 14% (air). He believed that seagulls, probably herring gulls, brought the crowberry to Surtsey as berries collected on Heimaey or, more likely, on Iceland, 30 km away. He found three plants of this small bush growing on flat ropy lava at the edge of the plant colony that had developed in the seagull nesting ground. The most recent plant to be brought to the gull colony area by seabirds, in 1998, was the mountain sorrel, *Oxyria digyna*, which is on Heimaey and common in Iceland ravines.

By comparison, on Rakata in 1908, 25 years after the Krakatau eruption, Docters van Leeuwen (1936) estimated that of the 110 vascular plants recorded by that time, only 23% were dispersed by animals (birds), 55% by sea and 22% by air. On Anak Krakatau by 1979, 26 years after the self-devastating 1952–53 eruptions, the percentages were: animals (birds, bats) 15%, sea 48%, wind 33%, human agency 4% (Table 12.3). Birds thus played a relatively more important role in the colonizing of Surtsey by vascular plants than they did on the Krakataus over equivalent time periods. The difference, of course, is due to the presence on Surtsey, but not on the tropical Krakataus, of seabird colonies, which are able to occupy an island before it can support resident land birds.

Migratory birds were almost certainly also involved in the colonization of Surtsey by plants. Of 230 species of migrant birds known in Iceland, 60 had been recorded on Surtsey by 1975. Redwing (*Turdus musica*), dunlin (*Calidris alpina*), snow bunting and pied oystercatcher (*Haematopus longirostris*) were seen on the new island as early as the spring of 1964. Unlike the stray migrants, some of the 'regulars', like ruddy turnstone (*Arenaria interpres*) and common redshank (*Tringa totanus*), arrived in flocks of hundreds of exhausted birds and, over the years, a merlin (*Falco columbarius*), long-eared owl (*Asio otus*) and short-eared owl (*Asio flammeus*) have been seen, exploiting these easy pickings.

Fridriksson's group inspected the feet of 90 individual migrating birds after they had landed on Surtsey. On 13 individuals they found such immigrants as diatoms, fungal hyphae, spores of moulds and moss spores. Over a period of 6 weeks in April/May 1967, 97 individuals of 14 species of migrating birds were captured, inspected for externally attached seeds, the gut flushed for contained seeds and the grit from the gizzard examined for its mineral and rock type as a clue to where the last intake of food may have occurred. The only birds found to be carrying seeds were snow buntings, of a subspecies that migrates from the British Isles to Greenland via Iceland.

Of 32 snow buntings captured, ten had seeds and grit in the gizzard (but no seeds in the stomach, indicating that no seeds were consumed a short time before capture). The seeds were of at least seven species of plants, most of them common in Iceland and the British Isles. Those of *Polygonum persicaria* and the sedge *Carex nigra* were germinated and grown to maturity by Fridriksson. One seed was of the bog rosemary, *Andromeda polifolia*, which is very rare in Iceland but occurs in Greenland and the British Isles. Others were possibly of *Medicago sativa*, a species absent from both Iceland and Greenland. Examination of grit from the gizzard showed it to be composed, not of old Icelandic basalts, but of metamorphic rock types and younger sediments that could not have been Icelandic, together with some Surtsey ash. The grit had not been acquired from Iceland, and Fridriksson cautiously concluded that both the seeds and grit were picked up by the birds either in the British Isles or (which seems much less likely) on Surtsey itself.

In April/May 1968, the project was repeated and 200 individuals of various migrating bird species were examined. This time there were only five snow buntings, and four of them had seeds in the gizzard. As in the previous year, seeds were not found in any other species. The seeds were of a species of *Silene* that grows in Iceland, and again the sedge *Carex nigra*. When they were tested for viability, one of the sedge seeds germinated and produced a seedling. Whether the seeds would have retained their viability after complete treatment in the gizzard and defaecation by the birds is not known. This time the gizzard grit was exclusively Surtsey tuff, but again there was no old Icelandic basalt. Although less conclusive than the 1967 results, the 1968 findings still showed that snow buntings carry seeds, and that seeds of two species of plants reached Surtsey, with or without the assistance of birds.

Taken together the results of both years strongly suggest that seeds of at least eight species of vascular plants were brought to Surtsey from the British Isles in the gizzards of snow buntings, probably directly, and the seeds of two species had retained their viability.

Earliest associations and communities of species

The first associations and communities of species have now been detected on Surtsey. In 1978 Fridriksson noticed that over a sandy area in the east of the island the sea sandwort, *Honkenya peploides*, a low-growing, sand-binding pioneer, was spreading into an area of lyme grass, *Elymus arenarius*, which had been present there for 4 years. Together with the lungwort, *Mertensia maritima*, these species appeared to be forming a simple sand-dune association, an association which, on Iceland, is succeeded by grasses. One cannot help noting the similarity between this association and that on the Krakataus between the beach-creepers *Ipomoea pes-caprae* and *Canavalia rosea*, and the grasses *Imperata cylindrica* (alang alang) or *Saccharum spontaneum* (glagah), although dunes are not found on the Krakataus. On Surtsey the sandwort is the pioneer, at ground level, holding in place a moist sand substrate, and the lyme grass grows above it.

The development of a dune in association with the tall lyme grass and the ground-covering sea sandwort was followed by Fridriksson in succeeding years. Sand from the beach drifts to the vegetated site and is trapped, gradually building up a dune. From a height of 20 cm in 1980, the dune had reached 1 m 10 years later (when I had the privilege of helping him measure it), and 135 cm in 1998, when 17 cm was added to the height in the last year; the base of the dune is now over 10 m wide. The dune has now begun to age; on its north-east side it has begun to erode. Its lyme grass plants have flowered every year since 1979, and the number of its flowering spikes increased continuously over the first decade of its life, making it the centre of origin for this species on the island.

Fridriksson also found a second link between these two species. In 1983 he noticed that young lyme grass plants were growing in the centre of some of the sandwort patches. In this situation the young grasses are better sheltered than on the open ground and receive a more even supply of moisture. He found a young lyme grass plant in each of 20 patches of sea sandwort, yet no new lyme grass seedlings were found in the open ground around the patches. In 1984 he found 100 such patches, one of them 350 m away from the nearest seed-bearing lyme grass plant. By 1986 there were 166 such patches. Thus by careful monitoring and observation Fridriksson discovered that the sea sandwort pioneer patch acts as a kind of nursery, providing favourable conditions for the establishment of a lyme grass plant within it.

Another 'embryonic' plant association involved three grass species (two of *Poa* and one of *Puccinellia*) and the low-growing, creeping perennial *Sagina procumbens*. A colony of *S. procumbens* had been developing since 1986 on the surface of ropy pahoehoe lava on the south of the island, where the *Sagina* and the grasses were well fertilized by the herring gulls breeding in this particular locality. The same association is found in Iceland on cliffs and seabird colonies, as well as around farm houses. Fridriksson reported that the soil in this seabird breeding area on Surtsey had 60 mg nitrogen and 7 mg potassium per 100 mg of dry matter, and 2.9 mg phosphorus in 100 ml of dry soil; in this place the soil had become quite fertile.

Fledglings of the greater black-backed gull were seen to be using the *Honkenya–Elymus* dunes for shelter in 1975 and 1976, and a pair of adults used as many as 20 sea sandwort plants as nesting material, together with the red fescue grass *Festuca rubra* and the sedge *Carex maritima*. In 1981 an aphid was abundant on the sea sandworts and a springtail was common in the rotting, fungus-covered branches of the larger clumps.

The rich moss vegetation supported several soil animals in 1976, but this fauna seems to have been destroyed by drought in 1978 and has only partially recovered.

Birds' nests also provide a habitat for land arthropods. In one nest of the greater black-backed gull were two species of staphylinid beetles, a springtail and a mite, and the decaying materials beneath the nest were exploited by larvae of four species of saprophagous and coprophagous flies. Two species of spiders, in their turn feeding on the insects, completed this community.

On the southern side of the lava apron another community developed, but at a pace much faster than the coastal association described above. On a flat surface a few procumbent pearlworts, *Sagina procumbens*, were joined by grasses, and soon there was a dense association of nitrophilous species. Over the past 14 years vegetation has spread rapidly and now covers some 6.3 ha, about 3.7% of Surtsey's dry land area, and is easily seen as a green patch from the air. It is in this rapidly growing community that most of the vascular plant colonists since 1990 (26 species in 8 years) have become established. Fridriksson's group examined pellets regurgitated by gulls at the breeding site, and found that the pellets contained about 10% by weight feathers, 3% sand, and of the remainder which was food, 80% was fish and 8% plant material, mostly leaves and hulls of lyme grass plants.

Thus after almost 30 years a simple and probably rather unstable ecosystem had developed on Surtsey, comprising a few species of vascular plants, seabirds, fungi and several species of land invertebrates. Two embryonic associations of plants and animals, one associated with slowly growing sand dunes, another with birds' nests, had been formed. Another, involving mosses, appears to have been badly damaged but is now showing signs of repair. Finally, in the past decade the activity and growth of the gull colony has resulted in a rapidly growing association of nitrophilous plant species at that site.

Differences from the colonization of the Krakataus

The colonization of Surtsey has been slower than that of either the 1883 Krakatau islands or of Anak Krakatau. Fridriksson showed that plants are capable of growing just as well on Surtsey's tephra as on other young volcanic tephra, provided that water is available, and concluded that the slow colonization of Surtsey is not due to the nature of its substrate. He removed tuff, sea sand and pumice from Surtsey to the Agricultural Experimental Station at Reykjavik on Iceland, and compared the colonization of these substrates with that of three similar areas, the substrates of which were: ash, pumice and cinders from the 1970 Hekla eruption (which occurred on Iceland); old sea sand; and peat. After three summers ten species had colonized the peat, and ten the old sea sand, but ten species had also colonized the combined Hekla 1970 substrates and nine, the combined Surtsey substrates; in fact three species had colonized the

Surtsey substrates only. The total plant cover on the three Surtsey substrates averaged 10%, compared to a cover of 6% for the old sea sand and 11% for the three Hekla substrates. These compared with 61% cover on the peat.

Surtsey's source pool of species, by comparison with Krakatau's, is an impoverished one, and the climate is much harsher. The early importance of mosses on Surtsey has no analogue on the Krakataus, and seabirds have been responsible for a greater proportion of zoochorous plant species colonizing Surtsey. The seabird colonies on Surtsey have provided a conduit to the island for energy from the surrounding sea to a much greater degree than has occurred on the Krakataus. As on Anak Krakatau, airborne invertebrates arrive on Surtsey in substantial numbers, and, also like the Anak Krakatau situation, Diptera form a major component of the airborne invertebrate fauna. An animal community exploiting this airborne fallout, as present on Anak Krakatau, Hawaii and other volcanic areas, has not been reported on Surtsey, however. A significant proportion of the invertebrates colonizing Surtsey were seaborne, and, as on Anak Krakatau, included collembolans and other insects exploiting washed-up carcasses.

Lessons from Iceland

Other island biologists may learn a great deal from the Surtsey investigations. Although the much slower colonization of Surtsey has enabled more finely grained surveys than were possible on the Krakataus, several aspects of the Surtsey work are worth emulating.

First, investigations started very soon after the eruption ceased. On Anak Krakatau, for example, investigations should begin immediately volcanic activity declines, and permits and other bureaucratic procedures, which are considerable, should be arranged well in advance so that movement to the islands can be made immediately volcanic activity permits, without further delay.

Second, surveys of Surtsey were regular and frequent (annual). Regular surveys at short intervals in the early stages of colonization and succession are crucial to enable the recognition of the very early associations (e.g. the dune and bird's nest associations on Surtsey), which are the components from which an island ecosystem is built. They also permit identification of the links involved in colonization (e.g. the raven's role in the dispersal of *Armeria*, and the part snow buntings played in bringing viable seeds to Surtsey). Such links are often more difficult to distinguish when colonization and succession have proceeded for some years. Early observations can obviate the need for later guesswork. Should yet another *tabula rasa* be produced on Anak Krakatau, for example, regular and frequent surveys should be made from the beginning. Moreover, these surveys should be extensive and carried out with precision. On Surtsey, it was possible to monitor individual plants for the first few years of the colonization, and such meticulous work provided insights into the way pioneer plant associations arose. On the small Tuluman Island such detailed mapping should still be possible.

The islands Anak Krakatau and Surtsey are similar physically, and work on both carries about the same degree of difficulty, both logistic and (differently) climatic. There is no reason, other than availability of time and modest funding, why a programme of the Surtsey type should not be followed on Tuluman and, at the next opportunity, on Anak Krakatau. The costs involved are low by modern research standards and the potential findings of considerable significance. On a cost-effective basis such research must be among the cheapest in the world. The work on Anak Krakatau of Tukirin Partomihardjo on plants and Darjono on birds, and on Long Island by Rose Singadan and Ruby Yamuna shows that appropriate expertise is available locally, and local biologists could form the nuclei of expeditions in the future to Krakatau, Tuluman and Long Island.

Finally, the Surtsey investigation was able to proceed in an environment undisturbed by human activity, apart from that of the scientists themselves. Although Anak Krakatau has become a significant tourist attraction in Indonesia, bringing welcome foreign currency to the country, this has not happened on Long Island. It is not too late to legislate to shelter Long Island, Motmot and Tuluman from tourist disturbance, and on the Krakataus it should be possible to reconcile the advantages of tourism and film crews with the requirements of research. I shall discuss this point further in Chapter 16.

CHAPTER THIRTEEN

Motmot: an emergent island in fresh water

Birth and physical development

Between 1969 and 1972, 50 km off New Guinea's north-east coast, there was an island in which there was a lake in which there was an island in which there was a lake. This intriguing Chinese-boxes system began during the 1968 eruption on Long Island.

In spite of its large caldera, Long Island was not recognized as an active volcano until 1943, when, on a US Army 1-inch-to-1-mile map of the island, a crescent-shaped shoal, not present in 1938, was shown just to the south of the middle of the caldera lake some 5 km from the nearest shore, and labelled 'active crater'. This must have been eroded away (winds on Lake Wisdom can be strong and waves high) because it was not to be seen in 1952. Between 1953 and 1955 intermittent Surtseyan activity was reported at the same site and a second island was formed, and there was probably a further eruption in 1961. By 1968 this island was represented only by a flat-topped islet 170 m by 70 m, a smaller rocky islet of scoria and boulders, and an even smaller fin-like spine 15 m long.

New activity was observed in 1968 (D'Addario 1972). On 19 and 20 March there was a disturbance in the sea at a point which had been the site of fumarol activity on the shoreline of the 1953–55 island, and a horseshoe-shaped crater, opening to the north, appeared above the water surface. By 25 March this subaerial ash cone had extended, to become fused with the spine and the flat-topped island and form the present island, Motmot (which simply means 'island' in the local language). On 29 March the temperature of lake water 2–3 km south-west of the centre of activity was about 30 °C. By 12 June there were two islands, and by November 1969 the crater was complete, with a crater pond, completing the system noted above of an island (Long) in which there was a lake (Wisdom) in which there was an island (Motmot) in which there was a small lake (the crater pond).

The two islands persisted until September 1970, by which time the second, smaller island had been reduced to a shoal and the crater pond was bubbling at various points (Johnson et al. 1972). Eighteen months later the crater had been breached by erosion. Fissures and a small pit were formed in 1972.

Eruptive episodes, probably in early 1973, produced a new ash cone in the north which was photographed on 27 April 1973, when weak activity was reported. On 1 and 2 May there was vigorous Surtseyan activity at both the 1973 and 1968 craters. By August, Motmot had grown in an easterly direction and by September, there was a new cone with a summit crater within the previous 1972 cone. On the eastern flank of this new cone a larger crater had white sublimates within and around its rim, suggesting that it may have been formed by a single phreatic explosion.

A glow was seen over Long Island from a passing ship in October 1972, and during activity in October and November six craters associated with the newest cone were active and an a'a lava flow formed a lava fan on the island's east coast. The surface of the north of Motmot was covered in scoria, including some bombs almost a metre across.

By February 1974 there had been further eruptions, including two small a'a lava flows, to the west and north-east, the latter reaching the shore, and the main cone had increased markedly in size. There was further activity between 23 and 28 February, forming another new cinder cone on the site of the first eastern lava fan, covering it. By now the island had basal ramparts of lava and its future existence was assured.

The colonization of Motmot

Motmot was investigated at intervals of 1 or 2 years from 1969 to 1978 by Ball and his colleagues (Bassot and Ball 1972, Ball and Glucksman 1975, 1978, 1980, 1981, Ball and Johnson 1976, Ball 1977, 1982a, 1982b), in 1988 by Osborne and Murphy (1989) and in 1999 by my own group of biologists.

Although Motmot is not particularly remote, as islands go (it is only some 4 km from the nearest shore of Long Island) its isolation is unusual. Certainly when spending the night on the island one gets a feeling of extreme isolation as one contemplates the great vault of the heavens covering the lake with its cliff ramparts in the distance all around, like some huge outdoor stadium. Motmot has a triple barrier. First, the marine barrier to Long Island, 50 km from the nearest New Guinea shore. Then the land barrier of Long Island itself – a ring of land that excludes all seaborne propagules from Motmot. Finally, Lake Wisdom, a freshwater barrier for both land and marine organisms, and a fairly good one even for fresh-water plants and animals (see the previous section). Lake Wisdom's sides are steep, with few shallow-water areas that favour aquatic life, and all Motmot's shores quickly fall off to the great depths of the lake some 350 m below. So, although close to the equator, Motmot gives one the impression more of Surtsey (of the impoverished source and harsh climate) than of Anak Krakatau, for example. There is no woodland, coverage by vegetation is sparse, and there are few birds. Even in areas of denser plant cover, most individual plants are less than 50 cm high. The plant assemblage (in 1999 comprising only 38 seed plants and seven fern species) was deemed 'species-poor' (Harrison *et al.* 2001) (Table 13.1). Long Island's flora is itself impoverished in relation to the adjacent Madang area of Papua New Guinea.

Table 13.1 *Comparison of the plant colonization of some small tropical volcanic islands*

	Size (km²)	Years from colonization	Dispersal barriers (km)			Number of species		Dispersal spectra (% species)			
			Sea[a]	Land	Lake	Ferns	Seed plants	Animal	Wind	Floating	Human
Motmot	0.1	26	55	4	4	7	38	42	44	13	0
Tuluman	0.3	20	300 (1)	0	0	6	53	39	20	39	2[c]
Anak Krakatau	3.1	27	44 (2)	0	0	4	60	17	22	59	2
Volcano Island	25.0	6	0	0	0[b]	21	258[d]	47	35	9	9

[a] Distance to continental source community: distance to nearest island in parentheses.
[b] Water is salty.
[c] At least ten species survived eruption.
[d] Eighteen species not identified.
Source: Harrison *et al.* (2001).

Lycosid spiders (now identified as *Geolycosa tongatabuensis*) and earwigs were found in high dry inland areas of Motmot in 1969, 'when, in spite of a careful search, no other life was found there' and were also seen hunting on the beaches. Ball and Glucksman (1975) found the role of earwigs and lycosids difficult to explain with certainty. Curiously, lycosid spiders were present very early in inland areas of the island where there was no obvious food for them. It was thought that chironomids (and earwigs) could be part of their diet, but lycosids can fast for more than 6 months. Ball and Glucksman suggested that a major part of the spiders' diet may have been the fallout of aerial plankton sucked in following updraughts. Lindroth *et al.* (1973) had suggested that an island may tend to receive more aerial fallout than its area would imply, because of a concentration of airborne material in its lee and updraughts on hot days sucking air towards the island from all directions. The latter point would apply particularly to tropical islands, and especially to those that are volcanically active. Damselflies and dragonflies, as well as Pacific swallows, *Hirundo tahitica*, may have exploited the aerial plankton before it reached the ground. All were present at the first survey, in 1969, and the swallow nested on Motmot by 1971. Together with earwigs, the lycosids gradually spread until by 1972 they could be found over almost the whole island.

On Surtsey, Lindroth and his colleagues had found that flies were an early group of colonists, dependent on carrion, and when there was no carrion they became extinct. On Motmot, scavenging and omnivorous earwigs were amongst the earliest arrivals, and were noted to swim rather well. They were probably not foraging on the beach, for they gradually moved inland. One earwig species, of the genus *Labidura*, was eating other earwigs and lycosids. There was a strand fauna of small beetles, ants and bugs, again depending directly or indirectly on the input of organic material from outside the island. The ants found among vegetation around Motmot's pond and staphylinid beetles and collembolans beneath a hardened algal crust at the pond margins had all arrived by 1972. The ants were surveyed by pitfall traps, and the ten species present are largely a subset of those on Long Island (Edwards and Thornton 2001), (Table 13.2).The most abundant species, *Anoplolepis gracilipes*, is a well-known invasive species, but had not been recorded from Motmot before 1999. In 1999 earwigs were again found on Motmot's east coast, where they occurred in low numbers under damp stones, but not in other areas.

The series of eruptions in 1973–74 enlarged Motmot, destroyed its crater pond and set the flora back to almost zero. After the eruptions only three individual plants (of *Cyperus cyperinus*) remained, but invertebrates survived, lycosids and earwigs still being abundant. When Eldon Ball returned to Motmot in 1980 and 1985 plant numbers were rising again and insect and spider diversity was high. The ferns *Pityrogramma calomelanos* and *Nephrolepis cordifolia* were widespread, and there were three species of figs.

Table 13.2 *Comparison of ant fauna (1999) on Motmot and on western lakeside fringe of Long Island*

Ant[a]	Motmot	Long Island
Anoplolepis gracilipes	w, (?qa), m	m
Camponotus quadriceps	qa	w, qa
C. novaehollandiae, s.l.		w
Camponotus sp.	qa	qa
Cardiocondyla minutior	w, qd	w
Iridomyrmex sp. (*anceps* group)	w	w
Leptogenys sp.		m
Monomorium sp.	w	
Odontomachus simillimus		w
Paratrechina sp.	w	w
Pheidole sp.	w, s	w
Polyrhachis (*Chariomyrma*) *gab*		w
Polyrhachis (*Hedomyrma*) sp.		w
Polyrhachis (*Cyrtomyrma*) sp.		w
Rhytidoponera araneoides		w
Technomyrmex albipes	qa	
Tetramorium sp. A	w, qd	w
Tetramorium sp. B		w
Turneria dahlii		w, qa
Formicinae sp.		m
Total worker diversity	7	14
Total ant species diversity	10	18

[a] Taxa represented by workers (presumed resident) denoted w, with others denoted as: a, alate; d, dealate; m, male; q, queen; s, soldier.
Source: From Edwards and Thornton (2001).

On Motmot, dispersal as a result of human activity has been minimal. From the composition and distribution of the flora, Ball and his colleagues believed that the plants were mostly derived from seeds carried by the black duck, *Anas superciliosa*, which nested on the island from 1969. Many of the plant genera on Motmot are among those found in the guts of black duck in the Northern Territory of Australia, although Frith (1982) thought it unlikely that seeds would survive passage through the gut of a duck. On Motmot, the ducks' gut contents on Long Island included many seeds of four dicotyledonous plants and one sedge, together with the shells of the freshwater mollusc *Melanoides tuberculatus* and insect remains (Kisokau 1974), but no seeds could be found in faeces examined. Thus seeds adhering to the birds externally, on feathers and in mud, are the more likely propagules. Osborne and Murphy (1989) pointed out that plants such as ferns, grasses and composites are more likely to be wind dispersed, and suggested that columbids were more likely than ducks to have played a role in the dispersal of figs and *Polygonum*, for example.

Kisokau (1974) examined the contents of the guts of six Long Island bird species, including two columbids. The gut contents of Stephan's dove, *Chalcophaps stephani*, and the grey imperial pigeon, *Ducula pistrinaria*, consisted entirely of unidentified fruit, grit and stones, and unidentified seeds were present in the gut of the dusky scrubfowl (or Melanesian megapode), *Megapodius freycinet*. The megapode is present on Motmot but no columbids have yet been recorded from there. The large imperial pigeons, species of *Ducula*, are obligate frugivores, feeding wholly on fruit. They generally do not digest seeds and they retain them in the gut for from 3 to 8 hours, so they are efficient dispersers of many forest trees. In Australia, *D. pistrinaria* has been found to void whole seeds of *Ficus, Terminalia, Solanum, Antidesma, Pandanus* and *Piper* (Frith 1982), all genera which have representatives on Long Island. On the Krakataus, the pied imperial pigeon, *Ducula bicolor*, which is also present on Long Island, is believed to be responsible for colonization of Krakatau by the palm *Coryphaca utan*. In contrast to species of *Ducula* (and fruit-doves of the genus *Ptilinopus*), the ground doves, *Chalcophaps* species, which take insects as well as fallen fruit and seeds, generally have a thick muscular gizzard that mechanically abrades seeds with gravel and grit, facilitating digestion. Thus ground doves can be regarded as seed predators rather than seed dispersers, although a few intact seeds have been collected from captive birds, so they may occasionally act as dispersers.

Ball and Glucksman thought that another likely mode of seed dispersal to Motmot could be within the prey that falcons carry to the island to feed. The two raptor species present on Motmot are the peregrine falcon, *Falco peregrinus*, and white-breasted sea eagle *Haliaetus leucogaster*. The lannar falcon, *Falco biarmicus*, is a second-hand disperser of plant seeds in this way in South Africa; from its regurgitated pellets, which included remains of a pigeon, a dove and a bulbul, no fewer than 15 plants, including two species of fig, were germinated (Hall 1987). There is (less conclusive) observational evidence that peregrines may act as dispersers in this way also on Anak Krakatau (Thornton 1994, 1996b, Thornton *et al.* 1996). On Motmot it has yet to be established that the falcons do or would carry avian prey from the lake shore at least 4 km over water to consume it on Motmot.

Along with a few invertebrates, some plant propagules arrive on Motmot via the surface of the lake. On a floating log with soil in its roots drifting past Motmot Ball and Glucksman (1981) found earwigs (not the species found on Motmot), ants (two species) and spiders (two species). However, the combination of rarity of waterplants, strong wave action, volcanically heated beaches and changing water levels ensures that surface transport is not a major contributor to Motmot's flora.

The number of species of vascular plants on Motmot increased from one (a single mature sedge plant) in 1969 to 12 in 1971 and 14 by 1972, when a total of 21 species had been recorded on the island. An eruptive episode in 1974–75 reduced this flora to three individuals of a single sedge species. The second succession proceeded, and by 1988 there were 19 vascular plant species, at least six of which had been part of the previous flora (Osborne and Murphy 1989): the fern *P. calomelanos*, the sedges *Cyperus polystachyos* and *C. javanicus*, *Ficus benjamina*, *Lindernia crustacea* and *Pipturus argenteus*. Plants present in 1988 but not previously recorded included the ferns *Nephrolepis cordifolia*, *Davallia solida*, and unidentified species of *Microsorium*, *Pteris* and *Blechnum*, and seed plants *Amaranthus cruentus*, *Galinsoga parviflora*, *Vernonia cinerea*, *Emilia sonchiflora*, *Imperata cylindrica*, *Ficus galberrima* and *F. opposita* and a species of *Dendrocnide* (Osborne and Murphy 1989).

Cyperus sedges were the most successful colonists. *Cyperus durides* was one of the few plants to successfully recolonize San Benedicto island after the eruption of Bárcena Volcano (see Chapter 5), a *Cyperus* species was one of the early plant colonists found by Kisokau on the emergent island Tuluman, in the West Pacific (below), and *Cyperus planifolia* featured in the early succession of Cayo Ahogado, a small sand cay off Puerto Rico studied by Heatwole and colleagues (below) (as did *Thespesia populnea* and *Ipomoea pes-caprae*, two sea-dispersed species that were important pioneers on Krakatau and Anak Krakatau). *Cyperus javanicus* and *C. polystachyos*, important pioneers of Motmot, also featured in the early colonization of Krakatau and Anak Krakatau. Sedges are well suited to the colonization of new soils, and Raab *et al.* (1999) have shown that 12 non-mycorrhizal sedges, including two species of *Cyperus*, utilize amino acids released during the decomposition of soil organic matter as a nitrogen source. They concluded that this capacity is widespread in the family Cyperaceae.

At the time of our expedition, in June and July 1999, at least 31 species of vascular plants were growing on the island, six species of ferns and 25 seed plant species, including eight *Ficus* species. Thus Motmot then had 19% of Long Island's known vascular plant species, 18% of its spermatophytes, and 26% of its *Ficus* species. There are few trees of stature; one plant each of a species of *Parkia*, *Alstonia* and *Dendrocnide*, and *Ficus benjamina* attains a height of from 5 to 10 m.

A great deal of turnover had taken place since the survey by Osborne and Murphy 11 years previously; 24 species (21 spermatophytes and three pterido-phytes) had arrived and become established and 11 species (eight spermato-phytes, three pteridophytes) had been lost from the island. New arrivals included bamboo, and noticeable individual trees, but still less than 10 m tall, were a specimen of *Alstonia scholaris*, a wind-dispersed tree, growing just within the high crater, a stinging tree, known as *salat*, also growing within the crater, and a specimen of a *Parkia* species, the seeds of which are wind-scattered, was

Table 13.3 *Dispersal modes of figs (* Ficus *spp.): numbers of species (a) on Long Island that have or have not reached Motmot in the first 31 years of Motmot's existence; (b) on the three older Krakatau islands that have or have not reached Anak Krakatau in the 46 years since devastating eruption of 1952–53; (c) that colonized the Krakataus in the first 25 years after the 1883 eruption*

	Dispersal mode[a]					
	ba	ba/bi	bi	bi/ba	?	Total
(a)						
On Long Island and on Motmot	3	1	3	1	0	8
On Long Island, not on Motmot	2	1	2	3	15	23
(b)						
On older islands and on Anak Krakatau	1	5	0	2	0	8
On older islands, not on Anak Krakatau	3	2	1	4	5	15
(c)						
On Krakataus in first 25 years	2	6	0	2	2	12

[a] Dispersal modes given as: ba, bat-dispersed; bi, bird-dispersed; ba/bi, primarily bat-dispersed; bi/ba, primarily bird-dispersed; ?, dispersal agents not known.
Source: From Thornton *et al.* (2001).

growing near the windward beach. Almost all the plant species were likely to have been dispersed by wind or animals, and about a fifth of Long Island's 167 (minimum) plant species, and the same proportion of its 31 fig species.

Two of the figs (*Ficus wassa* and *F. nodosa*) were fruiting at the time of our visit but their necessary and specific pollinating wasps had not yet arrived, for the figs contained no wasps or seeds. A free-standing specimen of *F. benjamina* growing just outside the crater was the largest tree on the island (about 10 m high) in 1999, and may have been more than 10 years old. It was large enough to be bearing a crop of thousands of figs, but was not fruiting. Two of Motmot's seven fig species are monoecious, five dioecious, about the same proportion (71%) as that on Long Island (61%). Only one animal on Motmot includes figs in its diet, the dusky scrubfowl (see below). Although breeding on the island, this incubating bird may be thought unlikely to be a very effective disperser, not being a frequent or strong flier, but it did fly off Motmot towards Long Island when disturbed during our visit.

Since there are virtually no fruit-bearing trees on Motmot, dispersal of figs to the island (Table 13.3) must now be a matter of chance overflights or touch-downs by effective frugivore dispersers that have eaten figs. From the fact that we frequently observed day-flying raptors, particularly the peregrine falcon, a specialist pigeon-feeder, it is likely that bats are more likely than frugivorous birds to set out across the lake on foraging excursions. Five of Motmot's seven fig species are known components of the diets of fruit bats, and the relatively greater importance of fruit bats over birds as fig dispersers in the early stages of colonization was noted on the Krakatau archipelago (Thornton *et al.* 1996).

A skeleton of an *Aplonis* starling, one of the probable fig dispersers on Long Island, was found with other bird remains on Motmot (see below), and it is possible that, as on Anak Krakatau, raptors occasionally bring fig seeds to the island in the bodies of their avian prey (Thornton 1994, Thornton *et al.* 1996). There is one other possible dispersal mode for some figs. Seeds of *F. benjamina*, one of the species now present, were found on the shore of Motmot in 1972, and whole figs of *F. virgata*, another species now present, were seen floating on the lake during that visit (Ball and Glucksman 1975).

Figs reached Motmot within 3 years of its emergence, and although eight species are now present, only two are bearing fruit, and these did not yet have their pollinating wasps in 1999 and did not set seed. It will be several years before the fig populations on Motmot build up to a size capable of supporting their own pollinator populations year round and thus become securely established on the island (Thornton *et al.* 1996). Studies on Anak Krakatau suggest that once one fig species reaches this stage of colonization, a feedback loop will begin to operate. There is likely to be a rapid increase in the visits of frugivores and the chances of dispersal of the other species of figs, as well as other fleshy-fruited plants, will be greatly enhanced. Once established, they in turn will enhance the island's attraction for frugivores, and so the feedback will proceed.

In all, in the two successions of plants since 1968, 57 vascular plant species have reached Motmot, 14 pteridophytes and 43 spermatophytes, and the two successions were almost entirely different in complement; only eight species being part of them both (42% of the 19 species present in the first succession, and only 14% of the total in both). This small proportion of species in common in the two early successions is in great contrast to the situation on Anak Krakatau, where a deterministic core of early successional species was recognized. This core, however, is made up of seaborne species, and there is no opportunity for these to play a role in the colonization of land-locked Motmot, although many of them are present on the shores of Long Island.

We found no amphibia, reptiles or ground mammals on Motmot in 1999, despite intensive search and the setting of traps. One bat species was living on the island, however: an insectivorous emballonurid microbat *Mosia nigrescens*, the lesser sheath-tailed bat. Three bird species were resident. The Pacific black duck and Pacific swallow were both still nesting there. There were at least six swallow nests on the low cliffs south of the landing beach and ten nests of the duck were found, most at the base of sedge or grass tussocks. The duck nests

contained clutches of from nine to 16 eggs but no ducklings were seen on
Motmot, although they were seen on the Long Island shores of the lake. It may
be that the ducks nest and lay on Motmot and when the eggs have hatched
safely, the brood is brought to the Long Island shores where more food is
available. The megapode, which Ball (personal communication) had seen in
1980 and found breeding in 1988, was still present and breeding in June 1999.
In New Britain this species lays eggs throughout the dry season (April to
December), with a peak in June–July (Jones *et al.* 1995). Other birds seen on
Motmot in 1999 include the rufous night-heron, *Nycticorax caledonicus*, white-
breasted sea eagle, *Haliaetus leucogaster*, Brahminy kite, *Haliastur indus*, and a pair
of peregrine falcons. Five duck skeletons, three of the collared kingfisher
(*Halcyon chloris*), and those of a singing starling (*Aplonis cantorides*) and
Melanesian megapode were found, evidently taken by raptors, most of them
probably by the falcons, as well as a white-breasted sea eagle skeleton.

On Lake Wisdom we observed red-throated little grebe (*Tachybaptus ruficollis*)
and Pacific reef-egret (*Egretta sacra*).

Most of the bird feeding guilds present on Long Island (see above) are absent or
poorly represented on Motmot – their niches are not yet available on the young
island. Rather, Motmot's avifauna is dominated by raptors, Pacific black duck
and Pacific swallows, which derive their sustenance from outside the island
itself (Schipper *et al.* 2001).

The raptors of course require territories very much larger than Motmot. For
example the sea eagle normally forages over about 13–40 km^2 and a pair cannot
survive on islands smaller than this (Diamond 1975). The kingfisher and starling
remains found on Motmot may be those of individuals killed there, or, more
likely, birds brought to Motmot by raptors to be consumed after being captured
on Long Island.

The Pacific swallow was one of the first birds to be recorded on Motmot (Ball
and Glucksman 1975). Other aerial insectivores are among the bird species that
may be expected to colonize the island in the near future. Ground insects are
scarce, but large wolf spiders are abundant all over the island. It is possible that
the scrubfowl, which normally feeds on ground insects and small fruit, takes
these. As the island's insects increase in numbers and diversity, feeding opportunities will also become available for Long Island's other insectivores.

Apart from the omnivorous scrubfowl, no frugivores have been recorded on
Motmot, which is not surprising considering the state of the vegetational succession and lack of fruiting trees. Until more trees mature and fruit is produced
the colonization of Motmot by additional frugivores seems unlikely. Similarly,
colonization by nectarivorous birds will be delayed until after the arrival of their
food sources. The grass and sedge seeds available on Motmot may begin to be
exploited by munias (*Lonchura* spp.), for example, before frugivores colonize,
although whether they would find sufficient sustenance now for a population to

become established is doubtful, and munias were not seen around the caldera rim. Perhaps the limited resources now available are in any case beyond the range of wandering finches.

Colonization by *Ficus* species

Figs reached Motmot within 3 years. Eight of the 11 species ever recorded there are dioecious, as are six of the eight now present. This proportion does not differ significantly from the proportions in the source faunas of Long Island, Madang Province or New Guinea (Table 13.3), and provides no evidence that either breeding system is advantageous in colonization. Six of the seven *Ficus* species on Anak Krakatau are dioecious, as are all seven species to reach the Krakataus within 25 years of the 1883 eruption.

The only fruit-eater on Motmot is the almost omnivorous Melanesian mega-pode, which has been recorded eating figs (W. R. J. Dekker, personal communication), so it could act as a disperser of *Ficus* species to Motmot and could be important in spreading them over the island when the immigrants have reached the stage of setting seed. Their numbers on Motmot are now low, however, and frugivores now have little reason to spend time on Motmot; its present fig flora is probably the legacy of frugivorous birds or bats occasionally visiting or overflying the island.

Raptors, probably largely the falcons, were taking a toll of ducks, kingfishers and megapodes on Motmot, and the starling skeleton was evidence of a frugi-vore and possible fig disperser either being killed on Motmot or brought there as prey. The presence of diurnal raptors may have an inhibitory effect on potential bird (but not bat) dispersers of figs; birds may be less likely than bats to fly over the lake on foraging excursions. Thus it might be expected that, as on the Krakataus, pioneer fig colonizers would tend to be bat-dispersed rather than bird-dispersed species. However a comparison of dispersal mode between those Long Island *Ficus* species that have reached Motmot and those that have not shows that bat-dispersed Long Island *Ficus* species are no more successful col-onists of Motmot than bird-dispersed ones.

There is evidence for the second-hand dispersal of *Ficus* species in Africa by a falcon, through passing fig seeds within pellets composed of its fruit-eating avian prey (Hall 1987). Also, on Anak Krakatau in 1992 seeds of the palm *Oncosperma tigillarium* were brought to the island in the body of a green imperial pigeon, *Ducula aenea*, killed by a raptor (Thornton 1994, 1996a). For at least the past 27 years peregrines have been using Motmot (Ball and Glucksman 1975, E. Ball, unpublished data), where they may well play a similar role. Until there is more plant cover, however, the presence of raptors is likely to deter potential bird colonists, particularly pigeons, as was believed to be the case on Ritter Island, also in the Bismarck archipelago, by Diamond (1974b) and on Anak Krakatau by Zann *et al.* (1990).

Table 13.4 *Cumulative numbers of species (colonization) by groups relevant in establishment of figs* (Ficus)*on (a) Anak Krakatau by 1992 following eradicating eruption in 1952 and (b) Motmot by 1999 following eradicating eruption in 1968*

(a) Anak Krakatau 1952–92								
Years since eruption	30	32	33	34	37	38	39	40
Raptors	1	1	2	3	3	3[a]	3	3
Fruit bats	1	1	2	2	2	2	2	4
Facultative avian frugivores	2	3	5	6	5	5	5	6
Specialist avian frugivores	0	0	0	1	1	1	2	2
Ficus spp.	2	2	2	2	3	5	6	7
Ficus spp., fruiting	0	0	2	2	2	3	3	5
Pollinating fig-wasp species	0	4	5	6	6	6	6	7
Other zoochores	5	7	7	7	10	13	15	15
(b) Motmot 1968–99								
Years since eruption	3	4	20	21				
Raptors[b]	3	4	?	3				
Frugivores	0	0	?	1				
Ficus spp.	2	2	5	8				
Ficus spp. with syconia	0	0	0	2[c]				
Pollinating fig-wasp species	0	0	0	0				
Other zoochorous plants	3	1	1	1				

[a] *Falco peregrinus replaced F. severus.*
[b] Not breeding on Motmot.
[c] Syconia unpollinated.
Source: After Thornton *et al.* (2001).

Ball and Glucksman (1975) found seeds of *F. benjamina* on Motmot's shore and figs of *F. virgata* floating on the lake, so one could theorize that some fig colonization may have been via the lake surface, although no fig specimens now grow at the shore, so that some secondary dispersal would have to be invoked.

Nearly all Motmot's fig plants are small and probably immature. Only when Motmot's fig trees mature and produce crops in such numbers that a population of their wasps can be maintained on the island year round, freeing them from dependence on airborne immigrant wasps, will the fig species' future be secure, frugivores attracted, and a feedback loop set in train that will accelerate the colonization of both figs and other fruit-bearing trees. On Anak Krakatau this stage was reached after 34 years (Thornton *et al.* 1996) (Table 13.4).

Motmot's food chain

In 1999 the fringe of sedges (*Cyperus polystachyos*) on the south-west margin of the overwash saddle supported a population of crickets (*Teleogryllus* sp.), which were also seen in the summit crater. Coccids, some attended by ants, occurred in association with black fungi growing on honeydew deposits on some fig trees and on *Ipomoea* stems. There was little other evidence of phytophage activity on trees, shrubs or ferns and we believe that the main source of energy for the island's food web is from outside the island – by air from Lake Wisdom and the surrounding ring of Long Island.

Adults of the caddis fly *Triplectides helvolus* were rising in numbers from the Motmot lake shore from just prior to dusk to at least 01:00 hrs. This caddis fly occurs in northern Australia and this is its first record from the New Guinea region. In 1999 chironomids of three species, as well as the caddis flies, were collected in a Malaise trap, in crawl-in tube traps (into which the chironomids were probably blown), and in nets set up at night. An hour after dusk (19:30 hrs) 14, 11 and 20 insects were counted passing from the lake through a vertical headlight beam in 1 minute; from midnight to 01:00 hrs there was still considerable activity – in ten counts an average of 5.5 insects were flying through the beam per minute (range 3–9).

A few damselflies, *Xiphiagrion cyanomelas*, had been noted in 1969 flying around the crater pond but had not been seen in subsequent visits to the island (Ball and Glucksman 1975). Damselflies, probably of this species, were common on the outer shore of Lake Wisdom in 1999 and were seen occasionally on Motmot. The dragonfly *Trapezostigma liberata liberata* was reported 'flying along some of the high ridges' especially at the south of the island in 1969, 'occasionally flying around the island' in 1971 and 'emerging in abundance' in 1972 (Ball and Glucksman 1975: 431). Large numbers of dragonflies emerged before and during our visit in 1999.

Tetragnathid spiders, *Tetragnatha nitens*, not reported previously, were abundant on Motmot in 1999. Present on almost every clump of sedge, by day they guarded egg cocoons fixed to the tips of leaves and at night took up station on silken threads spun between sedge plants and other prominences such as dead trees and projecting pieces of a'a lava. Damselflies and chironomids were commonly seen caught up on the strands.

Ground-hunting lycosid spiders, *Geolycosa tongatabuensis* (not *Varacosa tanna* or *V. papakula* (see Ball and Glucksman 1975)), were abundant, and at very high densities, from the shore to the island's summit crater, although, along with the earwigs, they had occupied a much-reduced area in 1976. They are nocturnal, and ten counts by spotlighting around midnight revealed *Geolycosa* densities averaging $13/m^2$ (range 7–22), smaller individuals occurring in the smoother, finer material of talus piles, larger ones where the cinders were over 1 cm in diameter. This spider is a shore species with a distribution extending from New Guinea through the south-western Pacific, including Vanuatu, Tonga, Samoa and French Polynesia to the northern tip of New Zealand.

There is a general paucity of ground insects on Motmot, and these dense spider populations and two ant species, including *Anoplolepis gracilipes*, probably exploit the fallout of emerging chironomids (four species), caddis and odonates on to the island's surface. It is likely that emergence flights from the lake at dusk and early evening constitute the main energetic input for the Motmot island food chain, being exploited in the air by day by swallows and dragonflies and by night by bats and, as fallout nears and reaches the ground by the two spider species.

The raptors, duck, swallow and emballonurid bat all appear to derive much of their sustenance from material originating outside Motmot.

The duck (and grebe) must be feeding largely on lake insects, for there is no aquatic vegetation. The swallow by day and the bat by night both capture chironomids and caddis flies emerging from the lake in large numbers. The swallow was one of the first birds seen on Motmot, in 1969 (Ball and Glucksman 1975) and, like the duck, is now an established breeder. The lesser sheath-tailed bat (one of the smallest of bats, weighing as little as 2.5 g), which also occurs on Crown and Umboi, uses acrobatic flight to forage within the understorey, capturing both flying insects and those clinging to foliage. We found it in the highest crater, where a bat had been seen and photographed in 1985 by Ball (personal communication) and was perhaps of this species.

It is possible that the megapode, which normally feeds on ground insects and small fruit, and perhaps even the night heron, *Nycticorax caledonicus*, may be feeding on the lycosid spiders. The Ashmoles speculated that, before the discovery of Ascension Island by the Portuguese in 1501, an endemic species of *Nycticorax* (now extinct) may have been an important predator in several animal communities, preying extensively on land crabs (*Geocarcinus lagostoma*) (Ashmole and Ashmole 1997). Another night heron, *Nycticorax violacea*, is believed by J. Llinas to prey on *Geocarcinus planatus* on Socorro Island, in the east Pacific Revillagigedos Group (Jimenez *et al.* 1994). However, if the heron and megapode are taking the wolf spiders on Motmot, they appear to be having little effect on numbers.

Absences from Motmot

It may be argued that if a species has made it to Long Island it should be able to reach Motmot, yet most of Long Island's bird guilds are absent or poorly represented on Motmot. There may be three main reasons for this. First, birds have had 350 years to reach Long but only 30 to get to Motmot. Second, the flight from Long Island to Motmot, although fairly short (4–8 km), is exposed, without cover, in the presence of raptors. For pigeons, white-eyes and finches the presence of falcons may be a sufficient deterrent. Third, and most important, the niches of many of these guilds are not yet available on the young island.

Ground insects were scarce on Motmot in 1999, and as the island's insects increase in numbers and diversity, feeding opportunities will become available for Long Island's insectivores. Colonization by nectarivorous birds will be delayed until after the arrival of their food sources. Similarly, considering the state of the vegetational succession and lack of fruiting trees (Harrison *et al.* 2001) it is no surprise that, apart from the omnivorous megapode, no frugivores have been recorded on Motmot. Until more trees mature and fruit, the colonization of Motmot by additional frugivores seems unlikely.

Collared kingfishers do not require trees for nesting or foraging. On Anak Krakatau they nest in the sides of ash-cliffed gullies and feed on shore crabs, and this kingfisher, and bee-eaters, which also nest in holes made in soft cliffs, may be early colonists of Motmot if sufficient food becomes available. Both are present on Long Island, and the kingfisher skeletons on Motmot may represent failed colonization attempts. As noted above, there is an abundance of food on Motmot for any animal immigrants capable of exploiting the dense population of fairly large spiders.

Birds such as munias may begin to exploit the grass and sedge seeds available on Motmot before frugivores are able to colonize, although it is doubtful that they would find sufficient sustenance now for a population to become established, and munias were not seen around the caldera rim (nor have they yet colonized Anak Krakatau). Perhaps the limited resources now available are in any case beyond the reach of wandering finches crossing open areas with little or no cover in the presence of falcons.

Aerial insects are the chief conduit for energy to Motmot (see above) and other aerial insectivores are among the bird species that may be expected to colonize the island in the near future and join the swallow, bat and dragonflies in exploiting the large numbers of aerial prey.

Generalist pioneers

Two of Motmot's vertebrate colonists exemplify the success of generalists as early colonizers of pristine habitats, as seen on the Krakataus (Thornton 1996a). Motmot's bat is a habitat generalist, having been found in habitats varying from primary forest to gardens and villages, and it may roost under the leaves of large-leaved plants, in caves or under roofs. The Melanesian megapode also is a generalist. It is practically omnivorous and has been recorded laying eggs on ground heated by volcanic activity or by the sun, as well as in mounds of vegetation, and has the widest selection of habitat for breeding site of any of the 22 species treated in Jones *et al.* (1995) and (p. 50) 'at the current level of knowledge this is the most diverse array of incubation techniques of any megapode'.

Comparisons

It is of interest to compare and contrast Motmot's colonization with the colonization of two other tropical marine emergent volcanic islands in the same region, Anak Krakatau, some 4500 km to the west, in Sunda Strait between Java and Sumatra, and Tuluman Island in the Admiralty Group of the Bismarck archipelago.

Table 13.5 *Numbers of species of* Ficus *arriving on the Krakataus during intervals between surveys, by reproductive habit, and by growth form*

	Period of arrival					On Anak Krakatau
	1883–1908	1909–22	1923–33	1934–51	1952–93	1992
Number of species	7	8	–	2	7	7
Dioecious	7	5	–	–	2	6
Monoecious	–	3	–	2	5	1
Large or medium tree	–	6	–	2	5	2
Shrub or small tree	7	2	–	–	2	4

Source: After Thornton *et al.* (1996).

Motmot and Anak Krakatau

Like Motmot, Anak Krakatau, is an emergent island volcano. There are, however, two major differences between the Anak Krakatau and Motmot cases. First, Anak Krakatau's volcanic activity has continued intermittently into the late 1990s. Second, Anak Krakatau is a marine island and can receive seaborne propagules and food items (flotsam); this important conduit for immigrants and energy is absent from Motmot.

On both islands there have been two successions. Motmot's first, starting in 1968, was destroyed by eruptions in 1972–73 and the second, from 1974, is still proceeding. In all, in the two successions, 59 vascular plant species have reached Motmot, 14 pteridophytes and 45 spermatophytes, and eight species of seed plants and one fern species were part of both (41% of the species present in the first succession). This is a smaller proportion than that common to the two successions on Anak Krakatau, where a deterministic core of early successional, widely distributed, largely seaborne plant species was recognized and 37 (86%) of the 43 species in the first succession were present in the second.

The first vertebrate colonists of Anak Krakatu were wading shorebirds (Zann *et al.* 1990), of course absent from Motmot. The early appearance on Anak Krakatau of raptors and aerial insectivores, birds that obtain food from resources not provided by the island, is a notable similarity between the two cases.

Anak Krakatau's pioneer fig trees first fruited 34 years after the 1952 eruptions and their fig wasps were already coming on to the island (Tables 13.3, 13.5). The fruiting coincided with an increase of frugivores present from four to seven, and this set off a colonization cascade. The number of fig species rose from two to seven in the next 5 years and the number of other zoochorous plants doubled (from seven to 15). Once figs became present in numbers and variety, specialist frugivorous birds colonized. This triggering effect may be expected also on

Table 13.6 *Numbers of species in common and (in brackets) similarity (percentage of smaller flora in common, and Sorenson coefficient) in vascular plant floras of Motmot in 1999, Anak Krakatau in 1983 and Tuluman Island in 1984; numbers of species on islands are shown after island names*

	Anak Krakatau (104)	Tuluman (42)
Motmot (57)	8 (14%) 0.10	2–5 (5–12%) 0.04–0.10
Anak Krakatau	–	22 (52%) 0.30

Motmot. Although smaller than Anak Krakatau and more than twice as far from fig sources, there is an abundance of excellent fig-dispersers and over 30 *Ficus* species available on the surrounding Long Island.

Motmot and Tuluman Island

Some 300 km north of Long Island, Tuluman emerged from the sea as a result of volcanic activity between 1954 and 1957 about 1 km south of Lou Island, itself south of Manus. Tuluman's colonization was monitored (Kisokau *et al.* 1984) some 30 years later, when the island was about 500 m in diameter, 23 m high and 0.28 km² in area.

At least 20 of Tuluman's 36 known seed plants are sea-dispersed. Fourteen of these were also part of Anak Krakatau's pioneer flora, along with two other sea-dispersed species, of *Casuarina* and *Timonius* (species unidentified on Tuluman). Tuluman's wind-dispersed grass, *Imperata cylindrica*, was also an important pioneer on Anak Krakatau. In all, the sequences on Tuluman and Anak Krakatau have at least 19 species in common, 45% of Tuluman's known flora (Table 13.6).

The sea-dispersed core of pioneers has no access to land-locked Motmot, although many of its component species are present along the sea coast of Long Island. In contrast to the comparison with Anak Krakatau, the sequences on Tuluman and Motmot have in common only the grass *I. cylindrica* and *Ficus opposita*. Five per cent of Tuluman's known flora (a Tuluman species of each of *Cyperus*, *Pipturus* and *Timonius* all with representatives on Motmot) are unidentified.

Only two reptile species have been reported from Tuluman, a crocodile and a monitor, most likely the same species as on Long Island, and neither of which has been reported from Motmot. Similarly, of Tuluman's nine resident land bird species, all except *Dupetor flavicollis*, the black bittern, were successful colonists of Long Island, but only one of them, the megapode *Megapodius eremita*, has also colonized Motmot (although there is evidence that one of the starling species reached there). The only mammal seen on Tuluman was a fruit bat, *Pteropus admiralitatum*. Two species of *Pteropus* have colonized Long Island, but the only bat living on Motmot is a very small emballonurid.

PART V

Colonization and assembly

CHAPTER FOURTEEN

Dispersal

Because of differential dispersal ability in the putative source biota, coloniza-tion rate and extent of any island is obviously influenced by the degree of isolation. Thus, the sea surrounding the Krakataus acts as a biological filter. At one extreme, it may assist dispersal. It is a highway for organisms like the mangrove *Lumnitzera* and many beach and coastal trees and shrubs, such as *Hibiscus*, *Terminalia* and *Casuarina*, that have buoyant seeds. The fact that the route to be traversed is over salt water is of no consequence in the case of small, light, fern spores or the minute balloon-like seeds of orchids, whose dispersal range is well above the distance to be covered. The stretch of sea is a partial barrier, of varying intensity, for some other anemochorous (air-dispersed) plant propagules, and for swimming or rafting animals like the large monitor, or geckos, rats and arboreal termites. For swimmers, of course, the travel distance may be an important limitation. Dispersal across Sunda Strait is also more likely in the case of some flying vertebrate species (and the organisms that they carry) and winged insects than others. The sea is virtually a complete barrier to species such as heavy-seeded primary forest trees, amphibians and large terrestrial mammals. The only amphibians known from Sebesi Island, between the Krakataus and Sumatra, are a frog, *Rana cancrivora*, and a toad, *Bufo melanostictus*, both of which are associated with people, with the first used as a source of food (there have been permanent inhabitants of Sebesi for some decades). The frog, a crab-eater, is also unusual in tolerating brackish water for breeding. Neither of these has colonized the uninhabited Krakataus, which lack significant bodies of fresh water. Thus the Krakataus' isolation, a sieve through which putative colonists must pass, is clearly the first factor controlling the assembly of a fauna there.

   The 1886 vascular flora comprised 16 species (62% of the flora present at that time, cf. data from Rakata alone, which were thought to have been dispersed by floating on the sea surface (thalassochory) and the rest were assessed as having arrived as seeds carried in the air (anemochory) (Treub 1888). Docters van Leeuwen (1936) considered that of the 64 species of the 1889 flora, 47% arrived by thalassochory, 44% by anemochory and 9% by zoochory (pro-pagules adhering to the outer surface or being carried within the alimentary

Table 14.1 *Dispersal modes of vascular plants recorded from Rakata at successive survey periods (some surveys grouped); numbers in parentheses are included pteridophytes*

Dispersal mode	Survey date					
	1886	1897	1908	1920–24	1929–34	1979–89
Sea	9	23	46	53	53	59
Wind	15 (11)	27 (13)	39 (17)	99 (51)	109 (54)	156 (81)
Animal	0	2	18	48	66	110
Human	0	0	0	18	8	4

Source: After Whittaker *et al.* (1992a) and Thornton (1996a).

tract of animals – mostly birds and bats). By E + 25 he considered that 60 species (52% of the 115-species vascular flora) had achieved a landfall by thalassochory, 32 (28%) by anemochory, and now 23 (20%) had arrived by zoochory. The sea-dispersed component had levelled off by 1928 (E + 45) and has since remained fairly stable, whereas the animal-dispersed component, which by 1934 (E + 51) had reached 68 species (a quarter of the 271 species then present), has become more important in recent decades (Whittaker *et al.* 1989) as more animal dispersers have become established (Tidemann *et al.* 1990, Thornton *et al.* 1993, 1996, Thornton 1994, Schedvin *et al.* 1995). Table 14.1 shows the changing dispersal spectrum of the various 'waves' (colonists in the interval between successive surveys) of the Krakatau flora calculated more recently on the basis of minimum turnover (Whittaker *et al.* 1989, Schmitt and Partomihardjo 1997), and differing somewhat from Docters van Leeuwen's figures.

In contrast to Docters van Leeuwen's assessment for plants, Dammerman (1948) considered that in 1908 (E + 25) 92% of the total fauna had been dispersed to the island group by air, in 1921 92%, and 1933 93%. My own assessment of post-1931 animal colonists would differ little from this. Clearly the sea has been and still is a far less important dispersal route for animals than it was for plants in the early decades. Only a few land lizards and snakes, land molluscs, insects such as termites and flightless soil-inhabiting invertebrates like earthworms, and possibly rats, have arrived by thalassochory.

Dispersal to the three 1883 Krakatau islands by anthropochory (through the agency of human activity) has been of little significance, most plant introductions having been shaded out by encroaching forest within a few years. The most important human-assisted animal arrivals were the rats: *Rattus rattus* on Rakata in 1920 after a family of settlers had lived there for 3 or 4 years, and *Rattus tiomanicus* on Panjang in 1928 after a visit by scientists of several months (although this species is known to be able to traverse 33 km of sea: Dammerman 1948).

Constraints on the dispersal of animals

The varying effectiveness of the marine barrier to dispersal may be highlighted by comparison of the Krakataus' complements of certain groups with representation of the same groups on the mainlands.

The first living thing to be found on the islands, 3 months after the 1883 eruption, was a spider – and there are now well over 100 species on the archipelago. Yet Bristowe (1931) remarked on the absence of spider groups that typically do not have immatures that disperse by 'ballooning' on silken threads. These include taxa such as mygalomorphs (including trapdoor spiders) and *Tetrastica* (apart from one species on Panjang that probably arrived by human agency). Indeed, the first mygalomorph was not found until 1990, on Sertung (W. Nentwig, personal communication), possibly having arrived on a floating log. The beach at the tip of Sertung's spit, now lost, was then a terminus on the islands for flotsam and was strewn with stranded logs.

No tree species typical of primary forest has colonized the archipelago. The powers of dispersal of primary forest trees tend to be poor – anemochory (as opposed to wind-scattering of seeds), for example, is of little adaptive value in forest interiors where species with heavy seeds, with large food stores, are at an advantage through their attractiveness to consumers.

Two species of land snails had colonized by the time of the first survey, 1908: the small litter species *Gastrocopta pediculus* and the larger arboricolous *Amphidromus porcellanus*. The latter was already numerous and present up to 370 m altitude (then the limit of collecting); it is now common in places on Rakata, and occurs on the summit, but has never been found on any other island. In the next intersurvey interval six more species colonized, all litter inhabitants. In the first 50 years 11 litter species and two arboreal species had arrived; in the second 50 years two litter and six arboreal species (Table 14.2). It is notable that not one of the five species, mostly large and conspicuous, recorded before the eruption, has recolonized, although 19 other species have done so. The 11 that are litter inhabitants arrived earlier, and now have a wider distribution among the Krakatau islands, than the eight arboreal species (nine litter and two arboreal species colonized within the first 50 years; six litter and no arboreal species now occur on more than two islands) and the correlation between early colonization and the litter habit is significant (Mann–Whitney z). Neither size nor ovovivipary appears to be correlated with colonizing ability, whether this is measured by the date of colonization of the archipelago or the number of islands colonized. Only one species has so far colonized Anak Krakatau, the small litter species *Liardetia doliolum*, the commonest and most widespread species on the archipelago.

Smith and Djajasasmita (1988) assumed that litter species are the more likely to be surface-dispersed and arboreal species the more likely to be wind-borne, and concluded that rafting was a more effective method of dispersal, even for

Table 14.2 *Krakatau land molluscs: incidence in surveys, and apparent habitat preferences*

Species	Incidence in survey						Habitat[a]
	pre 1883	1908	1919–21	1933–34	1982	1984–85	
Cyclophoridae							
Cyclophorus perdix	×						L
Ellobiidae							
Melampus flavus				×		×	L
Pythia chrysostoma			×	×		×	L
P. pantherina			×	×		×	
P. plicata			×	×		×	L
P. scarabaeus			×		×	×	
Veronicellidae							
Filicaulis bleekerii						×	L
Vertiginidae							
Gastrocopta pediculus		×	×	×		×	L
Succineidae							
Succinea minuta				×	×	×	L
Subulinidae							
Lamellaxis gracilis			×	×	×	×	L
Subulina octona				×	×	×	L
Valloniidae							
Pupisoma orcula						×	A
Achatinellidae							
Elasmias sundanum						×	A
Lamellidea subcylindrica						×	A
Helicarionidae							
Elaphroconcha bataviana	×						L?
E. javacensis	×						
Coneuplecta sitaliformis				×	×	×	A
Liardetia doliolum				×	×	×	L
L. indifferens			×	×		×	L
Microcystina gratilla						×	L
Camaenidae							
Amphidromus inversus	×						A?
A. banksi					×	×	A
A. porcellanus		×	×	×		×	A
Chloritis helicinoides	×						L
Landouria rotatoria						×	A
Pseudopartula arborascens					×		A
Total species	5	2	8	12	8	18	

[a] A, arboreal; L, litter.

Source: Smith and Djajasasmita (1988).

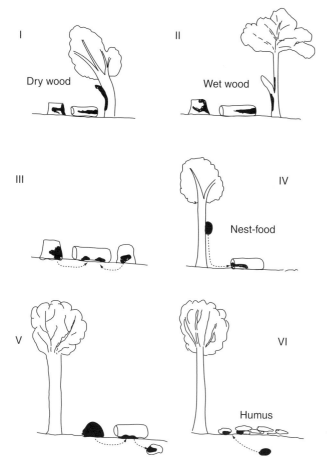

I

Dry wood

II

Wet wood

III

IV

Nest-food

V

VI

Humus

Figure 14.1 Diagrammatic representation of six lifestyle patterns of termites relevant to their colonization of the Krakatau Islands. Key to lifestyles: I, species nesting in and feeding on hard dry wood only; II, species nesting in and feeding on soft wet wood only; III, species nesting in soft wet wood and making galleries to some other wet wood; IV, species nesting on living tree trunks and making covered galleries to consume soft wood; V, species making mounds or subterranean nests and making galleries to consume soft wood and fallen leaves; VI, species nesting in soil and consuming humus. (After Abe 1984.)

very small species, than drifting on air currents. To further assess the long-term relative efficacy of these dispersal modes from the mainlands, one would need to compare the existing 11 : 8 litter : arboreal ratio found on the Krakataus with the ratio in the source areas, and the latter information is not yet available.

Such comparative data are available, however, for another animal group that has both ground-inhabiting and arboreal species – the termites, Isoptera (Figs. 14.1, 14.2, 14.3). The Krakatau termite fauna is disharmonic when compared to that of western Java (Abe 1984, Yamane *et al.* 1992). Species that nest in or on the surface of soil are notably lacking from the Krakataus, where species nesting in trees or in fallen wood predominate. This contrast would result if the fauna were accrued by the rafting of carton nests on driftwood rather than by flight. Scores of floating logs are found washed up annually on the Krakataus' high-energy beaches, and the maximum flight range of winged termites is thought to be only a few kilometres. Abe (1987) believed that dry-wood termites, which now dominate the Krakataus' termite fauna, are

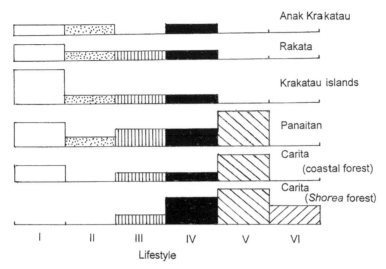

Figure 14.2 The lifestyles of termites on the Krakataus and other nearby localities; for lifestyle key, see Fig. 14.1. Vertical axis shows number of termite species in each lifestyle category, varying from one species (Anak Krakatau) to four species (Krakatau Islands) both category 1. (After Abe 1984.)

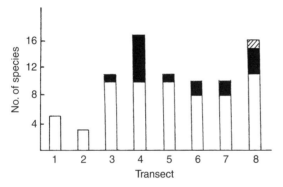

Figure 14.3 Numbers of species of termites found in transects on the Krakatau Islands and nearby areas. 1, Rakata; 2, Panjang; 3, Ujung Kulon, coast; 4, Ujung Kulon, inland; 5, Pangandaran, inland 1; 6, Pangandaran, inland 2; 7, Pangandaran; 8, Way Kambas. Open, wood-feeders; filled, soil-feeders; hatched, epiphyte-feeders. (After Gathorne-Hardy et al. 2000.)

particularly well pre-adapted for oversea dispersal. They have a coastal distribution, they are relatively tolerant of sea water, to which their nest wood is relatively resistant, and their caste differentiation is flexible.

Social bees and wasps that found their colonies by swarming are relatively rare on the Krakataus, and Yamane (1988) and Yamane et al. (1992) explained this by the inability of swarms to cohere over the sea distances involved. Of the three common species of bees of the genus *Apis* on the mainland of Java, one, *Apis dorsata*, is exceptional. Its swarms are unusually persistent over long distances (150–200 km), and this is the only one of the three species that has colonized the Krakataus. It had done so by 1919. Aculeate Hymenoptera exemplify well the difficulties of determining origins of the Krakatau fauna, but are markedly less diverse on the archipelago than on either Java or Sumatra (Tables 14.3, 14.4).

Table 14.3 *Turnover of the aculeate Hymenoptera fauna (excluding ants) on Rakata and Sertung together*

	1908	1919–22	1932–34	1982
Actual number of species (A)	17[a]	33[b]	41[b]	72
Number of 'new'species (B)	17	28	18	43
Number of species 'lost' (C)	0	12	10	12
First recorded species	17	28	14	37
Turnover rate[c]		80%	38%	49%

[a] One pompilid larva record excluded.
[b] Three undetermined Mutillidae excluded.
[c] Turnover rate between survey $(n-1)$ and survey (n) is $100(Bn + Cn)/(An - 1 + An)$.
Source: Yamane *et al.* (1992).

Table 14.4 *Numbers of species in selected groups of aculeate Hymenoptera known from Java, Sumatra and the Krakatau Islands*

	Java (J)	Sumatra (S)	J + S	Krakatau (K)	K/(J + S)	K/J
Vespidae	28	39	44	10	23%	36%
Swarm-founding social wasps	2	8	8	0	0	0
Trypoxylon	32	17	39	9	23	28
Bembecinus	7	?	>7	1	<14	14
Cerceris	19	?	>19	1	<5	5
Campsomeris, s.l.	23	16	26	5	19	22
Ceratina, s.l.	18	12	18	6	33	33

Source: Yamane *et al.* (1992).

The archipelago's complement of parasitoid braconid Hymenoptera is also biased in a way that reflects differences in dispersal ability that may have a behavioural basis. Koinobiont parasitoids do not kill the host on oviposition; their developing larvae feed within the host larva, emerging when the latter is mature. Idiobiont parasitoids, in contrast, kill or paralyse their host on oviposition. In the early braconid fauna of the Krakataus, and on the new island Anak Krakatau (Fig. 14.4), idiobionts are rather poorly represented and koinobiont parasitoids of Lepidoptera predominate (Table 14.5). The superior ability of the latter in colonizing Anak Krakatau may be explained partly by the relatively good representation on that island of their hosts, many of which are

Table 14.5 *Braconidae (Hymenoptera) from the Krakatau Islands, 1984–85, to indicate guild composition; note that differences in sampling effort (using beating and sweeping, Malaise traps, light traps, fruit-bait traps and pitfall traps) on different islands render direct comparisons of species richness unwise*

Island	Number of species	Ecological category[a]					
		ILD	IL	IC	KL	KD	KC
Rakata	47	0	6	8	14	14	4
Sertung	37	0	2	14	8	10	2
Panjang	36	1	2	12	10	7	4
Anak Krakatau	18	0	2	2	14	0	0
Total	102	Total I 38			Total K 64		

[a] ILD, idiobionts of Lepidoptera or Diptera; IL, idiobionts of Lepidoptera; IC, idiobionts of Coleoptera; KL, koinobionts of Lepidoptera; KD, koinobionts of cyclorrhaphous Diptera; KC, koinobionts of Coleoptera.
Source: After Maeto and Thornton (1993).

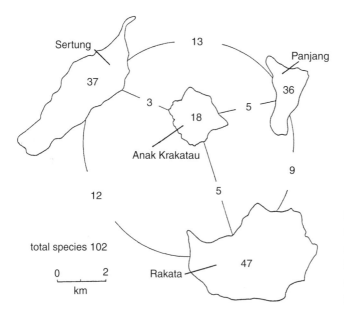

Figure 14.4 The braconid wasp fauna of the Krakatau Islands: the numbers of species found on each island and the number in common between each island pair. (After Maeto and Thornton 1993.)

early-colonizing, open-situation species, and partly by the fact that many koinobiont braconids themselves are thought to be *r*-selected species with high fecundity. It may also be that the generally oligophagous koinobionts suffered more than the more polyphagous idiobionts from the increasing fragmentation of host resources as diversity on the archipelago increased. If so, Anak

Krakatau's emergence may have provided a less host-diverse ecological refuge on the archipelago for such species. Another, more obvious, explanation is that occasionally koinobionts may be dispersed within their host larvae, whereas two separate colonization events are always required for the establishment of host–parasitoid systems involving idiobionts (Maeto and Thornton 1993).

The islands' butterflies include many well-known wide-ranging species, some having been seen in mass flights hundreds of kilometres from land (New *et al.* 1988, Thornton and New 1988b, New and Thornton 1992a). Four species of butterflies (one papilionid, two danaines and one nymphaline) were present in 1908, at the first zoological survey. The archipelago's present butterfly fauna is strongly disharmonic, heavily biased towards species with good powers of dispersal and wide distribution that feed on plants of secondary vegetation and transitional seral stages in coastal or near-coastal habitats. The flora of such habitats, in contrast to that of the interior, has changed little since 1897, and Anak Krakatau's flora is composed almost entirely of this unchanging floristic 'core'. Forest butterflies are probably less vagile and their larval food plants less likely to be encountered after arrival on the islands; there are few on the Krakataus.

Many dragonflies are strong fliers (Yukawa and Yamane 1985), and, like the butterflies, those occurring on the Krakataus are species that have been observed many miles from land. Water bodies for breeding have been small and scarce on the Krakataus, and although two libellulids were present as early as 1908 it is believed that few of the species recorded have been resident, breeding on the islands. Several species colonized in the 1930s when brackish lagoons existed on Sertung. Nymphs were found in three very small artificial ponds (Thornton and New 1988a), and no doubt some species can breed in small plant-held bodies of water. The species recorded on the islands are breeders in still-water bodies (van Tol 1990) and it is likely that there is regular monitoring of new habitats for breeding sites, those species with requirements for flowing water always being unsuccessful on the Krakataus.

The Krakataus' lack of amphibians must also be ascribed to dispersal difficulties, together with lack of habitat. Neither of the two species on Sebesi has been able to colonize the uninhabited Krakataus which lack significant bodies of fresh water.

The reptiles present are predominantly good swimmers (python, monitor) or species that are well adapted for raft transport (geckos, beach skink, blind snake). One rather unexpected absence is the flying lizard *Draco volans*, which is common in the mainland coastal areas. Several bird groups that are well represented on the mainlands and comprise or include poor or reluctant fliers such as babblers, pittas and phasianids are absent from the Krakataus, although the lack of appropriate forest habitat may also explain their absence (see below). Mammals are represented on the Krakataus by the two most vagile groups – rats and bats. Fruit bats colonized by the early 1920s (two species of *Cynopterus*), before the advent of microbats, probably because they are more wide-ranging (and generally have less specialized roosting requirements).

The frequency of arrivals

Long-term studies of the emergent island Anak Krakatau have provided some quite good data on this difficult topic.

Of course there is no direct way of estimating the frequency of visits to Anak Krakatau by individuals of those species that are already present. Such animals, of course, would not be identifiable as immigrants (in the absence of genetic analysis) since they could not be shown to differ in any way from residents. Individuals of species that do not occur on the island, however, are immediately recognizable as immigrants.

In June 1955 a dead hooded pitta, *Pitta sordida*, of the Sumatran subspecies, was found dead on the island (Mees 1986). More recently, a Danish bird observer, Torben Lund, reported seeing the Oriental white-eye, *Zosterops palpebrosa*, on the east foreland in August 1990, the only record of a white-eye on the archipelago. This inhabitant of open scrub occurs on Sebesi, and is a good island colonizer, but did not persist on the Krakataus. It is usually found in flocks, but the record was of a single bird only. In 1983 the pied triller (*Lalage nigra*), serpent-eagle, and a single leaf-warbler were first seen on the island. Sunda Strait is included in the range of two migrant species of leaf-warblers, neither of which has otherwise been recorded from the Krakataus.

We have seen other individuals which, like the pitta and leaf-warbler, were likely to be lone 'stragglers'. In 1986 a house crow, *Corvus splendens*, appeared at our campsite on Anak Krakatau and for two days frequented the camp kitchen, picking up scraps, and apparently unaffected by human activity. The bird was also seen to eat a fig. This species is a human commensal in the east Asian region and frequently travels with or on ships and boats. It appears to be extending its range eastwards through its association with marine commerce and competing, often successfully, with indigenous residents. The crow has not been recorded otherwise on the Krakataus and this individual was almost certainly a lone straggler. It may have come from a passing ship, as Sunda Strait is one of the world's busiest waterways. A tiger shrike was caught in a mist-net in the same year. Like the house crow, this migrant to the Indonesian area has not been recorded on the Krakataus either before or since. The chestnut-capped thrush (*Zoothera interpres*) is a bird of dense undergrowth which had been discovered on Rakata in 1984 and was found on Sertung in 1992. In 1986 a single individual was seen walking through the casuarina woodland of Anak Krakatau's east foreland; the species has not been recorded on the island since. In that year we also witnessed the arrival on the island, in daylight, of a Malay flying fox (*Pteropus vampyrus*), which stayed on the island for a day or so. It was probably a

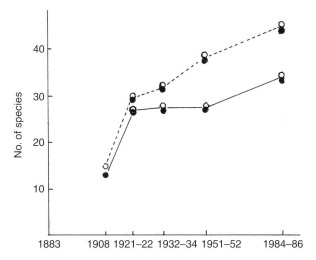

Figure 14.5 Colonization curves for land birds on Rakata. Solid line, actual curve; dashed line, cumulative curve, the difference between the lines representing putative extinctions. Solid circles, actual records; open circles, 'assessed records', including species that were earlier present, apparently absent at the time of the survey before that noted, and now present. (For details see Thornton *et al.* 1990a.)

straggler from the 'camp' of several hundred present on Sertung at that time. Very large fruit bats, probably of this species, have been disturbed near the summit of Rakata on two occasions in the 1980s and clearly bats often move around between islands. Our lucky discovery in 1990 of the tomb bat, *Taphozous longimanus*, was mentioned earlier. In the following year, after some practice on the mainland of Java, I used an ultrasound detector on the island each evening, but heard no calls of insectivorous bats. In 1992 we monitored ultrasound again, and again could detect no calls. The bat found in 1990 may have been a stray, but further investigation of the island by a bat specialist is now needed. In July 1992 a mature male of the rousette, *Rousettus leschenaultii*, the first record of this species on the archipelago, flew to fruit bait on the east foreland.

Since 1984 we have noted a number of arrivals of 'new' species to the island (Fig. 14.5), some of which resulted in successful colonizations, whilst others, although apparently not mere stragglers, arrived too recently for their status to be decided. For example, the distinctive calls of the Asian glossy starling (*Aplonis panayensis*) were heard on the north-east foreshore and the east foreland on our 1985 and 1986 visits but the species has not been recorded on the island since. The fact that birds were present in two successive years indicates that they (a pair was present in 1986) were not strays, but unsuccessful colonists that perhaps had arrived before sufficient fig trees were in fruit to sustain a breeding population. The barn owl, heard on the island in 1982, was evidently absent in 1984 but returned between 1985 and 1986. The Oriental hobby had been seen flying and hovering around Rakata's summit by Alain Compost in 1982; a pair was present on Anak Krakatau in 1986 and still present in 1989. In that year the hobby was supplanted by the peregrine falcon, and the falcon was still present in 1992. As noted earlier, the plain-throated sunbird (*Anthreptes malacensis*) probably became extinct on the island in 1984–85, but recolonized some time between 1986 and 1990. A pair and an immature of the pink-necked green pigeon were seen together regularly on the island in 1986, and two skeletons were found in the deep gully then known as Bone Alley. A pair and an immature of the red cuckoo-dove (*Macropygia phasianella*) were also seen several times in 1986. We have never observed the emerald dove (*Chalcophaps indica*) or the green imperial pigeon on the island, only their remains after apparently having been killed by raptors. The arrival between 1990 and 1991 of the black-naped fruit-dove (*Ptilinopus melanospila*) and its continued presence in 1992 probably indicates a colonization but the status of the species, as indeed of many species on Anak Krakatau following the recent series of eruptions, is doubtful. The discovery of two species of *Rattus* on the island in 1990 was mentioned earlier, and evidence of the presence of an otter in 1990 and 1991 in an earlier chapter. Finally, the fruit bat *Cynopterus brachyotis*, two species of munias, and the koel (*Eudynamys scolopacea*) were first recorded on the island in July 1992, and their present status is unknown.

No doubt there were many colonizations of the island by invertebrates during the past decade, most of which were unrecorded. In other cases there is good evidence of arrival. Two examples are the large black-and-yellow papilionid butterfly *Troides helena* between 1986 and 1989 and the antlion, *Myrmeleon frontalis*, between 1986 and 1990. Both are conspicuous as adults, being quite large insects, the large swallowtail being so strikingly coloured that even I, a colour-blind entomologist, could recognize it, and the antlion's presence is clearly signalled by the conical pits made by the larvae in ash or sand. Neither species was seen in 1984, 1985 or 1986, on all of which occasions a lacewing specialist (New) was present. Comparison of the butterfly surveys of New and his colleagues with those of the Bush and Yukawa groups has also shown that four skippers (Hesperiidae) of grassland (*Polytremis lubricans*, *Telicota augias*, *Hasora taminatus* and *Parnara guttatus*) arrived on the island in 1989 and another, *Pelopidas conjunctus*, in 1990 and there are many other examples. In fact if we take the conservative view that butterfly species first recorded on the island by Yukawa in 1982 (Yukawa 1984) and Bush in 1983 (Bush 1986) may all have arrived well before those dates, no fewer than 27 species have been discovered on the island since 1983 (New and Thornton 1992b). Eleven of these have not been found in surveys conducted since they were first seen and are presumed to have been stragglers.

Thus over a period of 8 years we have records of some 54 species of animals, 25 of them vertebrates, arriving on Anak Krakatau. Fourteen of these (the white-eye, two munias, leaf-warbler, house crow, tiger shrike, serpent-eagle, otter, rousette, tomb bat and four butterflies) were unknown on the other islands and thus were likely to be new arrivals on the archipelago. Over 20 of the 54 arrivals were stragglers that did not become established. Observers were present on

Anak Krakatau over this period for a total of only about 4 or 5 months. Now take into account the following: arrivals of individuals of species already present would have been undetectable by observers; unsuccessful or very brief unde-tected colonizations could have occurred during the more than 90 months when no observers were present; and comparative data are not yet available for many groups of invertebrates. It is clear that hundreds of individuals must have arrived on the island over the 8-year period, many of them being of species new to the island and many probably representing failed colonization attempts.

CHAPTER FIFTEEN

Stepping stone islands: the case of Sebesi

One way by which the chance of a propagule reaching a given island may be enhanced is by using an intermediate island as a 'stepping stone', so in landscape ecology terminology increasing the 'connectivity' between different habitat 'patches'. The theory of stepping stone function was addressed by MacArthur and Wilson (1967), who devoted a chapter of their seminal book to the topic, and by Gilpin (1980). In 1996 I gathered an international team to examine the actual case of Sebesi Island in Sunda Strait, with a view to assessing its importance as a stepping stone in the colonization of Krakatau.

A number of hypotheses, both equilibrial and non-equilibrial, have been proposed to explain variation in island biodiversity (see Whittaker 1998). What is more difficult, however, is to predict which species will be represented in particular island ecosystems. In tackling this problem, Diamond (1975) derived assembly rules for the avifaunas of islands of the Bismarck archipelago based on the identification of exclusion pairs and the incidence functions of species, representing their varied abilities to withstand competition in avifaunas of various sizes. In the build-up of communities, contemporary competition theory implies an effect of priority, early arrivers excluding later competitors, and Ward and Thornton (2000) showed that pioneer communities should be more predictable than later-successional immigrants. The presence of island stepping stones is theoretically likely to influence both the size and make-up of isolated biotas.

The theoretical importance of stepping stones

Theoretically, the presence or absence of a stepping stone island between a major source population and an uncolonized 'target' island may have important ecological and evolutionary effects on a target island biota by influencing both the rate of colonization and the composition of colonists.

A stepping stone is usually smaller in both size and habitat range than its source, and, together with its own isolation, these constraints mean that it can harbour only a sample of the source biota for possible colonization of the target. But if the target has about the same size and range of habitats as the stepping stone, then, by reducing its isolation, a stepping stone may be important. Its most obvious effect would be to increase the number of species, especially amongst the poorer dispersers, that reach the target. It may also increase a species' chances of establishment by boosting the number of arriving individuals or by facilitating a rescue effect (Brown and Kodric-Brown 1977). The stepping stone may affect priority on the target, which, in turn, may partly determine the course and nature of colonization through inhibition (e.g. by competition, predation, habitat modification) or facilitation (e.g. through food provision, habitat modification, pollination). A stepping stone may also affect the future evolution of isolated populations by influencing the make-up and size of founder groups.

MacArthur and Wilson noted that a stepping stone's importance increases rapidly with decreasing dispersal power of the organism concerned and can quickly become paramount (providing that there is some dispersal). After accounting for the known mean overland dispersal distances of various species of plants and animals they were led to conclude that 'dispersal across gaps of more than a few kilometres is by stepping stones wherever habitable stepping stones of even the smallest size exist' (MacArthur and Wilson 1967: 133). Gilpin (1980) pointed out that a stepping stone's effect will be greatest when the distance between source and target is such that species' chances of reaching the target without stepping stone assistance are close to neither 1 (such as for many seaborne plants or fruit bats in the Krakatau situation) nor 0 (as for rhinoceros or many primary forest trees such as dipterocarps), but intermediate.

MacArthur and Wilson (1967) calculated that, other things being equal, for organisms with an exponential dispersal distribution the probability of reaching a target from a source with a stepping stone would be twice that of reaching it from a source without one. This would apply to the propagules of good passive dispersers with average dispersal range much greater than the distances involved – many microorganisms, small light-bodied insects, fern spores, the seeds of orchids, very small, light, dust seeds of other plants, plumed seeds that are wind-dispersed rather than merely wind-scattered, and floating propagules of many plants highly adapted for sea dispersal. In contrast, for organisms with a normal dispersal distribution, the probability of reaching the target from a source with a stepping stone would be a hundred times that of reaching it from a source without one. These would include rafters like geckos, rats, some snakes, arthropods and plants, in which the tendency to sink increases with time and progressive waterlogging, and swimmers like monitor lizards, pythons and (again) rats, as well as flying animals such as large-bodied, strongly flying insects, birds, and some bats, which tire at an increasing rate with time (and of course any plant or animal propagules they carry). In theory, when there are two stepping stones, interchange between source and target becomes polarized (MacArthur and Wilson 1967). Two stepping stones near a source increase its unaided contribution to the target by a factor of almost 5, but only double the reverse flow. Gilpin (1980) also pointed out that species incidence on a stepping stone will be boosted by the presence of other stepping stones between it and the source.

In spite of these contributions to stepping stone theory, there is still no consensus about how important stepping stones might be identified or indeed about the importance of any particular stepping stone in the development of any particular island biota.

Here we examine the possibility of stepping stone function in practice by considering the role of Sebesi Island in one of the best-known island colonization situations, that of the recolonization of the Krakatau archipelago following the 1883 explosive eruption of Krakatau Island, in Sunda Strait. Several researchers have speculated on the extent to which Sebesi, situated halfway between Krakatau and the rich continental source island of Sumatra, may have served, and may still serve, as a stepping stone, either for particular species (e.g. Dammerman 1948, Rawlinson *et al.* 1990, 1992, Whittaker *et al.* 1997) or more generally (Hoogerwerf 1953a, Thornton 1996a, 1996b, Partomihardjo 1997, Whittaker *et al.* 1997, Runciman *et al.* 1998, Yukawa *et al.* 2000).

The Sunda Strait islands

Recall that the Krakatau archipelago, in Sunda Strait, is some 44 km from the nearest shore of each of the large, biotically rich, continental islands of Sumatra and Java (Fig. 8.1, p. 121). There are four Krakatau islands. Three (Rakata, Sertung and Panjang) were effectively sterilized by the devastating 1883 eruption of Krakatau Island, of which Rakata is the remnant. The fourth, Anak Krakatau, emerged from Krakatau's submarine caldera in 1930 and is an active volcano. Sebesi and Sebuku islands lie between the Krakataus and Sumatra. Sebuku is a low eroded volcano with about the same area as Sebesi, lying between Sebesi and Sumatra. It was severely affected by the 1883 tsunamis and has since been cultivated. No native forest remains. Sebesi, about 2.5 km south of Sebuku and 16 km from Sumatra, is the closest land to the Krakataus, being 16 km from the nearest Krakatau coast. It is about 7 km in diameter and on the north and east a coastal plain extends some 1 or 2 km inland (Fig. 15.1). Sebesi's height and area (about 844 m and 30 km^2) are of the same order as those of Rakata (about 790 m and 12 km^2). Sebesi's lowland vegetation was severely affected in 1883 by ash deposition and tsunamis, but not to the same extent as Krakatau. Coconut plantations now extend to the uplands, where coffee is grown. Scattered remnants of natural vegetation remain between plantations but continuous forest cover is now restricted to an area above about 600 m (Partomihardjo 1999).

The strongest prevailing winds in Sunda Strait, the north-west monsoon, average about 28 km/hr but sometimes reach 45 km/hr. They would tend to bring airborne propagules to the Krakataus from Sumatra and Sebesi rather than from Java, although they predominate in the wetter season (late November

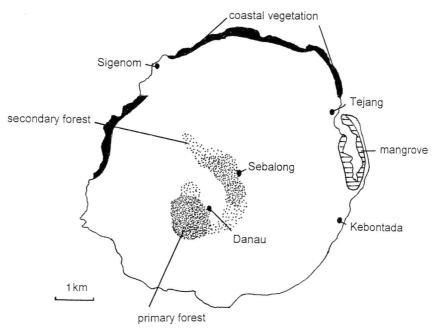

Figure 15.1 Sebesi, a possible stepping stone island for colonization of the Krakatau Islands.

to February), when some animals such as butterflies are less likely to be airborne and available for wind assistance. In March, winds are variable but becoming more southerly, and in the drier south-east monsoon season (April to October), winds are predominantly from Java. They swing to west-south-west in November, heralding the return of the north-west monsoon.

The surface of the Java Sea is several centimetres higher than that of the Indian Ocean and the strongest prevailing sea currents in Sunda Strait are those flowing from the north or north-east for 18 hours of the day during the south-east monsoon, with surface velocities of 0.38–0.83 m/s (Gathorne-Hardy *et al.* 2000). In November the flow reverses for 2 or 3 months, but this current rarely exceeds 0.25 m/s. Thus the prevailing flow, from the Java Sea to the Indian Ocean, would favour movement of floating propagules from Sumatra to Sebesi to Krakatau rather than the reverse.

Thus on theoretical grounds Sebesi may be expected to have had a significant stepping stone role in the colonization of the Krakataus from Sumatra, at least for some organisms. It is in the right position, is about the same size and height as the target, and the predominating air and sea currents are favourable. If its role were important, what testable predictions would follow?

Types of evidence of stepping stone function

Three types of evidence may provide insights into the stepping stone problem. The first is independent of colonization of the stepping stone but the others are based on the assumption that a species using an island in this way will first colonize it and then proceed to colonize the target. It is conceivable that an animal species may use an island literally as a stepping stone, perhaps extending its range thither for feeding, being unable to breed there and colonize yet dispersing further to the target, where conditions may permit colonization. The usual course of events, however, and all cases involving plants, will entail prior colonization of the stepping stone.

(1) Predominance of Sumatran rather than Javan forms on Krakatau

With source biotas about equally rich, more species would be expected to colonize the target from the source that is served by a stepping stone than from the source that is not. Although Sumatra and Java are almost equidistant from Krakatau, no island intervenes between Krakatau and the nearest coast of Java. If Sebesi has had a significant stepping stone role it should have facilitated colonization from Sumatra and, *ceteris paribus*, the Krakatau biota should contain a preponderance of Sumatran, rather than Javan, immigrants.

Dammerman (1948) believed that most of Krakatau's fauna was of Sumatran origin, but hard evidence is scanty and equivocal. In the absence of genetic information, only when a species has distinguishable Javan and Sumatran subspecies or is known to be absent from either Java or Sumatra can the source be determined. Most animal species occur on both sources with populations on each at present indistinguishable – although future genetic studies may elucidate the origins more precisely. Moreover, coverage of western Java and southern Sumatra by zoologists has been unequal and the extent of the disparity varies between animal groups. Accordingly, lack of records may indicate absence more reliably on Java than Sumatra and this may have greater force in some groups than others.

For non-migrant land birds, data from the nearest national parks in Java (Ujung Kulon, about 60 km to the south of Krakatau) and Sumatra (Bukit Barisan Selatan, some 180 km to the north-west) can be compared with data from Sebesi and Krakatau. Of the 267 species present in one or both of these parks (Thornton *et al.* 1990a, 1993, O'Brien and Kinnaird 1996), 216 (81%) have failed to colonize Sebesi, 220 (82%) have failed to colonize Krakatau and 206 (77%) have colonized neither. Fifty (35%) of the 142 bird species that occur in both parks have reached at least one of the islands, 26 (18%) reaching both, eight Sebesi only and five Krakatau only. Of relevance in this context, of 23 Ujung Kulon species that are unrecorded in Bukit Barisan Selatan, 19 (83%) have reached the islands; nine (39%) have reached both, five Sebesi only and five Krakatau only. In contrast, only one of the 102 Bukit Barisan Selatan species that are not known from Ujung Kulon has reached Sebesi and none has reached Krakatau. These comparisons certainly do not support a Sebesi stepping stone hypothesis.

The bat *Hipposideros cineraceus* is known from Sumatra, the adjacent Riao archipelago, southern Borneo, and, surprisingly, Kangean Island north-east of the north-eastern corner of Java, but not from Java itself (Corbet and Hill 1991). Two Krakatau bats are known only from Java: *Rhinolophus celebensis javanicus* and *Glischropus javanus*. The latter was discovered on Krakatau in the 1996 survey reported here, having been known previously only from a record on Mt Pangranggo, West Java, some 60 years before. This strongly high-flying species may have escaped detection on Krakatau, where high-canopy mist-nets have been used only rarely. The species was detected in 1996 through its call, and males and lactating females were trapped at the small Sertung spring and others netted over the small waterholes on Rakata. The evidence from bats is enigmatic.

Even taking note of the caveat of the Dutch workers Dammerman (1948) and Toxopeus (1950) that the distribution of some Javan forms of, for example, butterflies and birds may extend into the Lampung district of southern Sumatra, in general, distributional data provide no evidence of a preponderance on Krakatau of taxa with a clearly Sumatran, as opposed to Javan, origin. Rather, the reverse seems to occur.

The question is not simply one of preponderance of taxa of Javan or Sumatran origin, for the source faunas of the group in question may not be the same size, and the question becomes: has a greater proportion of identifiable Sumatran (not Javan) forms colonized Krakatau than of identifiable Javan (not Sumatran) ones? Recent surveys of termites on Krakatau, Sebesi, Java (Ujung Kulon and Pangandaran Nature Reserve in south-eastern West Java Province) and Sumatra (Way Kambas National Park) by Gathorne-Hardy and colleagues enable this question to be addressed. Nineteen species were present on Java but not Sumatra, and five of these (19%) were present on Krakatau, whereas of 18 species present on Sumatra but not Java, two (11%) were present on Krakatau (Gathorne-Hardy *et al.* 2000) The differences are small and neither support nor oppose a stepping stone role.

DNA analysis of populations of carefully selected species of plants and animals are now needed to test the hypothesis advanced at the beginning of this section, and such studies are beginning.

(2) Sequential records indicate movement from Sebesi

If the stepping stone has been effective, distributional evidence may be expected to show movement of species from stepping stone to target island rather than the reverse. Whittaker *et al.* (1997) pointed out that one might expect to find that a substantial number of species recorded on Sebesi but not on Krakatau in 1921 were found on Krakatau in later surveys, and the reverse should not be the case. However, without any stepping stone function, it is likely that Sumatran species would colonize the closer, less devastated Sebesi, before they colonized the more distant Krakatau with its thicker ash deposits. So this test would need to provide unequivocal, strong data.

The historical distributional data for plants, land molluscs, butterflies, reptiles and birds permit this test.

Whittaker *et al.* (1992a) cited 19 species of vascular plants (three pteridophytes, 16 seed-plants) found on Sebesi in 1921 but on Krakatau only after 1936. They noted, however, that this was a very small proportion (actually 5%) of Sebesi's 1921 flora, that many other species show the reverse sequence, and that the Sebesi data are poor, with no data corresponding to pre-1921 Krakatau surveys. They concluded that the Sebesi stepping stone hypothesis cannot be tested by this means.

Later, Whittaker *et al.* (1997) noted that three wind-dispersed trees on Krakatau, *Pterospermum javanicum* (found in 1992), *Alstonia scholaris* (1979 as a large specimen) and *Oroxylum indicum* (1989), were present on Sebesi in 1921. The first has fairly large winged seeds generally regarded as being merely locally scattered by wind, the second has plumed seeds capable of being carried over fairly long distances, and the third has large flattened seeds enclosed in a very light, large membranous wing. *Oroxylum indicum* also has a strict establishment constraint – the presence of a particular species of pollinating bat, the cave fruit bat, *Eonycteris spelaea*. This bat has not yet been recorded on either Krakatau or Sebesi, although the plant found on Krakatau in 1989 had almost certainly been pollinated for it was a mature tree with seedlings nearby; the bat is probably present but so far undetected. Whittaker *et al.* (1997) believed it possible that these three tree species colonized Krakatau from Sebesi.

Nineteen pteridophyte and 203 spermatophyte species (excluding cultivated species) were known from Sebesi but not from Krakatau in 1921 (the date of Sebesi's last survey before the 1980s). Of these, nine pteridophyte and 51 spermatophyte species were recorded from Krakatau in later surveys, 47% and 25%, respectively, of the species available to colonize it. However, ten (53%) of the 19 pteridophytes and 152 (75%) of the 203 spermatophytes have still not colonized Krakatau.

There is even less evidence of movement from Krakatau to Sebesi. Only three (8%) of the 37 pteridophyte species and seven (7%) of the 103 non-cultivated seed plants found on Krakatau by 1922 and not found in the 1921 Sebesi survey were subsequently recorded from Sebesi. This polarization gives some support for a stepping stone function for Sebesi.

Considering two groups of characteristically wind-dispersed plants, orchids and pteridophytes, of 29 species present on Sebesi but not Krakatau in 1921, half (15) were found on Krakatau subsequently. In contrast, in 1921 Sebesi carried 15 animal-dispersed *Ficus* species, five of which were not then established on Krakatau. One of these five, *F. asperiuscula*, had been recorded from Krakatau (Sertung) in 1920 but was regarded as having become extinct. It was

Table 15.1 *Diversity and affinities of butterflies recorded from Sebesi/Sebuko and the Krakatau islands*

Distributional categories	Recorded Sebesi/Sebuko	Recorded Krakatau
Java	6	3
Sumatra	9	2
Total	15	5
Total species recorded	52	98

Source: Yukawa *et al.* (2000).

rediscovered on Rakata and Panjang in 1995, possibly having recolonized (Whittaker *et al.* 2000). Two others, *F. drupacea* and *F. virens* (both bird-dispersed), were not recorded on Krakatau until 1989 and the remaining two, *F. crininervia* and *F. pilosa*, have not yet been found there. Thus three of Sebesi's five potential 1921 fig colonists to Krakatau were successful, in a period when 11 additional volant frugivores colonized Krakatau. Four more *Ficus* species were not recorded on Sebesi until the 1989–96 surveys; one was also a late colonist to Krakatau and the others are yet unknown there.

There were 16 species of non-marine molluscs on Sebesi in 1921 (two of them freshwater species), 14 of which were then unknown on Krakatau. Since 1921, although 15 mollusc species have been first recorded from Krakatau, only three of them were previously known from Sebesi, a land snail, a slug and a freshwater snail. Thus there is evidence of possible sequential movement of just three of 14 species (21%). Land molluscs were not surveyed on Sebesi in 1996 so we have no indication of the reverse movement.

Twenty-one of Sebesi's 1921 butterfly species were then unknown from Krakatau and seven subsequently colonized. However, eight species recorded on Krakatau in 1921 or earlier were not found on Sebesi in 1921 but found in 1994, that is, as cases of possible movement from Krakatau to Sebesi, and Yukawa *et al.* (2000) concluded that the 'evidence suggests that there have been more frequent movements of butterfly species from Java to the Krakataus and Sebesi–Sebuku, and then to southern Sumatra rather than in the opposite direction' (Table 15.1).

Seven reptile species were recorded on Sebesi by 1921, three of them already known from Krakatau. Of the four potential colonists, two, the king gecko (*Gekko monarchus*) and paradise tree snake, had reached Krakatau by the early 1980s, and two, Bowring's skink and the Malayan racer, have not yet been recorded. Only the records of the common house gecko (*Hemiductylus frenatus*) suggest movement from Krakatau (recorded 1908) to Sebesi (first recorded 1955) but

Table 15.2 *Resident land birds recorded from Sebesi and the Krakatau islands during various survey intervals*

	Sebesi		Krakatau islands						
	1921	1996	1908	1919–21	1932–34	1951–52	1984–86	1989–92	1996
Total species in survey	31	49	16	28	30	36	38	36	36
Cumulative total	31	51	16	30	34	40	45	47	47
Turnover									
Gains		20		14	4	6	5	2	0
Losses		2		2	2	0	3	4	0

Source: Based on Runciman *et al.* (1998).

the presence of a permanent human population on Sebesi for at least nine decades makes this difficult to accept.

Of seven bird species present on Sebesi in 1921 that had not colonized Krakatau, only two have since done so. A third, the spice finch (*Lonchura punctulata*), reached Krakatau in 1994 but did not become established. These three species could have arrived on Krakatau via Sebesi. Six species that were recorded on Sebesi only in the 1996 survey, and presumably reached Sebesi after 1921, cannot be taken as indicating movement from Krakatau to Sebesi for they were also relatively late arrivals on Krakatau (Table 15.2). Two species were present on Krakatau but not Sebesi on or before 1921 and first recorded from Sebesi in 1996: *Pycnonotus aurigaster* and *Copsychus saularis*. The former became extinct on Krakatau after 1908, so could not have moved to Sebesi from Krakatau; the latter is the only probable case of movement of a bird species from Krakatau to Sebesi.

In sum, evidence of movement of species from Sebesi to Krakatau is weak. Sequential records, over a spectrum of biotic groups and for a substantial proportion of potential Krakatau colonizers, is lacking. Only in pteridophytes and reptiles did about half the available species subsequently colonize Krakatau (47% and 50%, respectively). Bearing in mind the caveat made above, Thornton *et al.* (2002) did not regard the 21% of available land molluscs, 25% of seed plants, 29% of resident land birds or 33% of butterflies as sufficiently high proportions to support the hypothesis.

(3) The biota of Krakatau should be a subset of that of Sebesi

Assuming that Sebesi functions as a stepping stone only for species that have first colonized it, and, further, that such species persist there until the time of a survey (neither of which may be true, although both are likely to be), then some limits to stepping stone function can be indicated. The Jaccard index of similarity between stepping stone and target (J, the proportion of the combined biota that is common to both $= C/T$, where C is the number of species in common and T the total number of species in the two areas) provides an upper limit: the proportion of species that *may* have reached Krakatau via Sebesi.

Three of four factors listed by Gilpin and Diamond (1982) as contributing to the non-random association of species pairs on islands may also contribute to similarity between island faunas that have been assembled independently: shared habitat, shared geographical origin and shared distributional strategy. Shared habitat between Sebesi and Krakatau has been limited and not well synchronized (see below) but it is likely that some vagile species from the same source and/or with the same distributional strategy, for example with a predilection for islands of a certain size or with a biota of a certain size, were more likely to arrive on both islands independently than species chosen at random. If there has also been stepping stone assistance for less vagile species, the similarity will be further increased but this increase will be difficult to detect, let alone measure. A low similarity coefficient, however, would indicate poor stepping stone success.

More telling than degree of similarity is the extent to which the target island's biota is a subset of that of the stepping stone. If stepping stone function has been important, significantly more (and a greater proportion of) Sebesi species would be expected to be absent from Krakatau (SK_0) than Krakatau species absent from Sebesi (KS_0). However, although differences in survey coverage between Krakatau and Sebesi would tend to result in $KS_0 > SK_0$, giving a Krakatau > Sebesi nestedness pattern, a Sebesi > Krakatau pattern ($SK_0 > KS_0$) would result without stepping stone function. Sebesi, the closer to Sumatra, would be likely to have both good and poorer dispersers, whereas poorer dispersers would be likely to be lacking on Krakatau. Since Sebesi > Krakatau nestedness may be expected for this reason, a measure of nestedness will be an overestimate as an indication of stepping stone function. Low nestedness, therefore, would be a strong indicator that the stepping stone has been unimportant.

We considered evidence available from surveys of plants, land molluscs, termites, butterflies, amphibia, reptiles, non-migrant land birds, rats and bats (Table 15.3).

We counted the total non-cultivated seed plant species known from Sebesi as 425, from Krakatau 447, the overlap as 175 species and the combined known flora as 697 species (Whittaker *et al.* 1989, 1992a, Partomihardjo 1995). Twenty-two percent of Sebesi's pteridophytes (10 of 46) and 41% of its spermatophytes (175 of 425) are not known from Krakatau. More than half (272 species, 61%) of Krakatau's spermatophyte flora is not known from Sebesi and may be presumed to have colonized Krakatau without using Sebesi as a stepping stone. Moreover, many of the Krakatau species that are known from Sebesi may not have colonized Krakatau from there. The Jaccard index for spermatophytes is low (Table 15.4),

Table 15.3 *Numbers of bat species from Sebesi, Krakatau and adjacent areas of Sumatra and Java, with indication of dispersal ability based on foraging strategy (×, present)*

Species	Dispersal category[a]	Sebesi	Krakatau	Java	Sumatra
Cynopterus brachyotis	IV	×	×	×	×
C. horsfieldii	IV		×	×	×
C. sphinx	IV	×	×	×	×
C. titthaecheilus	IV	×	×	×	×
Macroglossus minimus	IV		×	×	
M. sobrinus	IV	×	×	×	×
Pteropus vampyrus	V	?×	×	×	×
Rousettus amplexicaudatus	V	×	×	×	×
R. leschenaultii	V	×	×	×	×
Emballoneura monticola	III		×	×	×
Taphozous longimanus	III		×	×	×
Megaderma spasma	I	×	×	×	×
Rhinolophus celebensis	I		×	×	
R. pusillus	I		×	×	×
Hipposideros cineraceus	I	×	×		×
H. diadema	III		×	×	×
H. larvatus	I	×	×	×	×
Glischropus javanus	II		×	×	
Kerivoula hardwickei	I		×	×	×
Murina suilla	I		×	×	
Myotis muricola	II	×	×	×	×
Scotophilus kuhlii	III		×	×	×
Chaerephon plicata	III	×	×	×	×
Total		12	23	22	19

[a] Dispersal categories I, II are microchiropterans with energetically expensive slow flight for short-distance foraging (I are poorer dispersers than II); IV are megachiropterans with fast manoeuvrable flight over short distances; III (microchiropterans) and V (megachiropterans) have fast prolonged flight enabling long foraging journeys. Thus strategies I, II and IV are poorer dispersers, and strategies III and V are good dispersers: see McKenzie *et al.* (1995).
Source: After Thornton *et al.* (2001).

and there are slightly fewer, and a slightly smaller proportion, of SK_0 species than KS_0 species, and thus little support for the stepping stone hypothesis.

Since 1883, 20 species of non-marine molluscs have been recorded from Krakatau (Smith and Djajasasmita 1988) and 17 from Sebesi (Dammerman 1922, 1948), with six species in common. It is noteworthy that none of the five species present on Krakatau before the 1883 eruption has recolonized, nor has any of them been recorded from Sebesi. In 1921 there were more than three

Table 15.4 *Jaccard and Sorenson similarity indices for selected biotic groups between Sebesi and the Krakataus and Sebesi and Panaitan*

	Jaccard	Sorenson
Sebesi/Krakataus		
Seed plants	30	46
Butterflies	33	47
Land molluscs	28	44
Reptiles	57	73
Birds	55	71
Bats	42	59
Total vertebrates	52	68
All above animals	44	61
All above biota	34	50
Panaitan/Krakataus		
Birds	55	71

times as many species on Sebesi than on Krakatau – 16 compared to five (Dammerman 1948). Two of those on Sebesi that had not reached Krakatau (*Phaedusa sumatrana* and *Pupina superba*) were unknown from Java, and Dammerman suggested that they may be expected to colonize Krakatau soon. Evidently they have not done so; neither was recorded from Krakatau by mollusc specialists in 1982 (Yamane and Tomiyama 1986), 1984 (Smith and Djajasasmita 1988) or 1996 (B. Scott, unpublished data). The human-dispersed giant African snail, *Achatina fulica*, has colonized Sebesi since Dammerman's surveys (it was found there in 1996), but has not reached Krakatau. Two relatively large arboreal snails that colonized Krakatau between 1933 and 1982, *Amphidromus banksi* and *Pseudopartula arborascens*, are known only from Panaitan, not from Java, Sumatra or Sebesi. The proportion of KS_0 species (11 species, 65% of the fauna) is about the same as the proportion of SK_0 species (14 species, 70%). Had Sebesi acted as a significant stepping stone one would expect the differences in these proportions to be greater. The Jaccard index of similarity for the molluscan fauna of Sebesi and Krakatau is low, 28%.

Nine of Krakatau's 12 termite species are unknown on Sebesi, and ten of Sebesi's 13 species have not been recorded from Krakatau; of the joint fauna of 21 species, only three are in common. There is no evidence of a stepping stone role from the comparison of KS_0 and SK_0 termite species. At least 75% of Krakatau's species have probably arrived without colonizing Sebesi, and about the same proportion of Sebesi species have not reached Krakatau.

Some 51 (65%) of the 79 butterfly species recorded from Krakatau in the period 1984–93 were KS_0 species, as they had not been recorded from Sebesi by 1993

Table 15.5 *Amphibians and reptiles recorded from Sebesi in 1921 and 1996, and on the Krakataus, with date of first record on the Krakataus;* × *denotes recorded*

| | Sebesi | | First Krakatau record |
	1921	1996	
Coluber melanurus	×	–	–
Emoia bowringi	×	–	–
Bufo melanostictus	–	×	–
Rana cancrivora	–	×	–
Python reticulatus	×	–	1908
Lepidodactylus lugubris	×	–	1920
Chrysopelea paradisi	×	–	1982
Gekko monarchus	×	–	1984
Varanus salvator	×	×	1889
Hemidactylus frenatus	–	×	1908
Mabouia multifasciata	–	×	1924
Gekko gecko	–	×	1982
Emoia atrocostata	–	–	1919–33
Ramphotyphlops braminus	–	–	1984
Hemiphyllodactylus typus	–	–	1985
Cosymbotus platyurus	–	–	1928 only

Source: Data from Dammerman (1922, 1948), Rawlinson *et al.* (1990, 1992) and Thornton *et al.* (2001).

(Yukawa *et al.* 2000). Sixty-three (64%) of the 98 species ever known from Krakatau are KS_0, 36 (51%) of the 71 species known from Sebesi are SK_0. Had Sebesi served as a stepping stone for most of the fauna, a smaller proportion of KS_0 than SK_0 species would be expected.

Of the 12 reptile species known from Krakatau, five (42%) are KS_0 species (Table 15.5): the skink *Emoia atrocostata* (on Krakatau from 1919 to 1933, now extinct there), the geckos *Gekko gecko*, *Hemiphyllodactylus typus* and (found in 1928 only) *Cosymbotus platyurus* and the blind snake *Ramphotyphlops braminus*, a small, highly vagile burrowing snake often human-dispersed, which may well be present on Sebesi. There is about the same proportion of reptile SK_0 species (four of the 11, 36%). *Elaphe flavolineata* (the common Malay racer) is a small terrestrial or semi-arboreal constrictor of the more open parts of wet tropical forests, open dry forests, savannas, grasslands and plantations, and is often found close to human habitations. It feeds preferentially on small mammals, particularly rats, and occasionally birds. Juveniles also take lizards and amphibians. On Sumatra it is found from sea level to 900 m and it also occurs on Java. This snake appears to be a good candidate for colonization of Krakatau but has not yet been found there. Other SK_0 reptiles are the tree snake *Dendrelaphe ?pictus*, which would have a competitor on Krakatau in the paradise tree snake, *Chrysopelia paradisi* (not found on Sebesi in 1996), the skink *Lygosoma bowringi*, which is often associated with humans and was thought to have been human-dispersed to Christmas Island (Cogger 2000) and the house gecko *Gehyra mutilata*. Once again, it is not the case that there are more (or a greater proportion of) SK_0 than KS_0 species. The Jaccard index of similarity for reptiles is 57%.

Twenty-nine of the 31 species making up Sebesi's 1921 resident non-marine avifauna were recorded again in 1996 (Runciman *et al.* 1998). The two that were not are the Asian koel, *Eudynamys scolopacea*, and the plaintive cuckoo, *Cacomantis merulinus*, and the latter has not reached Krakatau. Of the 51 species known from Sebesi, 14 have not reached Krakatau and two more have reached there without becoming established. Thirty-five species of the joint fauna of 63 are common to the two; Sebesi has been ineffective as a stepping stone for almost half (44%) of the joint avifauna. The numbers and proportions of SK_0 (16, 31%) and KS_0 (12, 26%) species are about the same.

There is no evidence for a stepping stone role in the case of rats. Krakatau's two rat species include one of Sebesi's two species, *Rattus rattus*, and not the other. The presence of this rat (and perhaps also *Rattus tiomanicus*) on Krakatau can be ascribed to human activity (Rawlinson *et al.* 1990).

All 12 of Sebesi's bat species are included in the known bat fauna of Krakatau (20 species) (Table 15.3), so rather than the fauna of Krakatau being a subset of that of Sebesi, the reverse is the case. Although all Sebesi species are known from Krakatau, the Jaccard index is fairly low (42%: Table 15.4) because, on the evidence, at least eight species colonized Krakatau without colonizing Sebesi. Differences between the known bat faunas, however, may simply reflect the much greater coverage of Krakatau, for Sebesi was not visited by bat specialists until 1996. In any case, the flight range, particularly of the fruit bats, is such as to make a stepping stone unnecessary.

In sum, in neither animals nor plants, taken as whole groups, is the Krakatau biota a subset of that of Sebesi; only in amphibia, land birds and zoochorous plants is SK_0 greater than KS_0. The Jaccard index of similarity between Sebesi and Krakatau, taken over the 1116 species used in this analysis, is not high (34%: Table 15.4). However, assuming that colonization from a pool of species on Sumatra is to some degree deterministic (Ward and Thornton 2000), one would expect considerable commonality between the resulting faunas on Sebesi and Krakatau, even without the involvement of a stepping stone. The extent of similarity from this cause may be expected to be lower, the poorer the colonizing power of the group concerned. The order of decreasing Jaccard indices between Sebesi and Krakatau for various biotic groups supports this to some extent (although reptiles may perhaps be expected to be nearer the bottom of the list): bats, birds, *Ficus* species, reptiles, rats, zoochorous seed plants,

Table 15.6 *Numbers of seed plant, butterfly and land vertebrate (excluding amphibia) species ever recorded on the potential stepping stone islands Sebesi and Panaitan and recorded on the Krakatau islands, and the numbers and percentages in common between these possible sources and the Krakataus*

	Seb	Also K	%	K	Also Seb	%	B	C	C/B%
Sebesi									
Seed plants	265	143	54	385	143	37	507	143	28
Butterflies	26	22	85	63	22	35	67	22	33
Reptiles	10	8	80	12	8	66	14	8	57
Birds 1	48	29	60	35	29	83	54	29	54
Birds 2	52	35	67	47	35	74	64	35	55
Bats 1	11	10	91	15	10	67	16	10	67
Bats 2	11	10	91	23	10	43	24	10	42
Total 1	95	69	73	125	69	55	151	69	46
Total 2	99	75	76	145	75	52	169	75	44
	Pan	Also K	%	K	Also Pan	%	D	E	E/D%
Panaitan									
Birds	66	40	61	47	40	85	73	40	55

Species known from the Krakataus by only a single individual are not included. 1, based on species as present in 1996; 2, based on cumulative records; K, Krakataus; Pan, Panaitan; Seb, Sebesi; B, number of species on Sebesi and the Krakataus taken together; C, number of species common to Sebesi and the Krakataus; D, number of species on the Krakataus and Panaitan taken together; E, number of species common to the Krakataus and Panaitan.
Source: Data on seed plants from Whittaker *et al.* (1992a), on butterflies from Bush *et al.* (1990) and Yukawa *et al.* (2000).

pteridophytes, butterflies, sea-dispersed Leguminosae, spermatophytes, land molluscs, termites, amphibians (Table 15.6).

In an attempt to make some estimate of the extent of commonality that might be ascribed to the deterministic element of colonization, reference was made to the proportion of the biota that Krakatau shares with an island in the south of Sunda Strait that is likely to have had a much less important stepping stone role – Panaitan (Fig. 8.1, p. 121).

The comparison with Panaitan

The great tsunamis of Krakatau's 1883 eruption devastated Panaitan's lowlands, but extensive upland areas were unaffected, and ash fall was very much less than on Sebesi (Hommel 1987, 1990). Throughout the period of Krakatau's recolonization Panaitan has been little disturbed by human activity, and is close to a relatively undisturbed region of Java (Ujung Kulon National Park, of which it is a part). By contrast, Sebesi has been disturbed considerably by humans (see below) and the adjacent areas of Sumatra have been increasingly developed for agriculture, forestry and urbanization. About four times the area of Sebesi, Panaitan is less than 10 km from Java, and, as well as good dispersers, may carry rather more of the poorer dispersers than does Sebesi, which is 15 km from Sumatra. The poorer dispersers on Panaitan, however, are less likely to reach Krakatau, some 45 km away, than are those on Sebesi, some 15 km from Krakatau.

Deterministic colonization patterns are likely to be the main reasons for similarity between the biotas of Panaitan and Krakatau. However, because the difference in isolation from the mainland is greater for the Panaitan/Krakatau comparison than for Sebesi/Krakatau, the commonality between Krakatau and Panaitan is likely to be less than that between Krakatau and Sebesi, regardless of stepping stone function. A higher similarity for Krakatau/Sebesi would therefore not be good evidence of a stepping stone role. Equal similarity, or greater similarity for Krakatau/Panaitan, however, would indicate little or no stepping stone role for Sebesi. Only the data for butterflies and birds allow us to test this.

Twenty-eight butterfly species have been recorded from both Panaitan and Krakatau (Table 15.1), 29% of Krakatau's 98 species, 97% of Panaitan's 29 species and 28% of the combined fauna of 99 species. The percentage of shared species with Krakatau is considerably higher for Panaitan (97%) than it is for Sebesi (49%) and the percentage of Krakatau's butterfly fauna shared with Sebesi (36%) is little higher than that shared with the more distant Panaitan (28%). There is little difference in Jaccard index. These comparisons appear to indicate that Sebesi has been unimportant as a stepping stone for butterflies but are surely affected by the fact that Panaitan is under-surveyed. Yukawa, the only specialist to collect there, could spend only 3 days in 1982 (Yukawa 1984, Yukawa *et al.* 2000).

Although the birds of reef, shore and secondary forest overlap considerably between Krakatau, Panaitan and Sebesi, Krakatau has no species that cannot live outside primary monsoon or rainforest, habitats that it lacks. Eight species that inhabit primary forests on Panaitan and in Ujung Kulon have colonized Krakatau, but have done so only recently and none is strictly confined to such forest (Zann *et al.* 1990). Similarly, the absence from Krakatau of most of Panaitan's stork-like and heron-like birds and kingfishers has been explained by lack of appropriate freshwater habitats. Lack of suitable habitat is not a satisfactory explanation for all other cases of absence, however. Hoogerwerf (1953b) listed ten Panaitan bird species that had not colonized Krakatau by 1951 although appropriate habitat appeared to be available there, and he predicted that in time these species would colonize. None had done so 45 years later.

Thirty-three non-migrant land bird species are common to Panaitan and Krakatau, 70% of Krakatau's 47 species and 41% of Panaitan's 81. Even excluding 15 species of herons, storks and kingfishers, 50% of Panaitan's avifauna is absent from Krakatau compared to 31% of Sebesi's 51 species. As expected, Jaccard's index is lower for the Krakatau/Panaitan comparison (35–41%, depending on the above exclusions) than for Krakatau/Sebesi (56%). The avifaunal comparisons, therefore, provide no evidence against the hypothesis of a Sebesi stepping stone role, which is neither refuted nor supported by the necessarily limited comparisons with Panaitan.

Overall, although Sebesi's size, height and position intuitively suggest a stepping stone hypothesis, four predictions following from the hypothesis are not borne out by the available evidence. Why?

Above we considered a measure which combined two sources of dissimilarity: species found on target but not on putative stepping stone, and species found on stepping stone but not on target. It is important to consider these separately, and their relationship to one another.

At the limit, if Sebesi's stepping stone role was paramount (i.e. Krakatau received no species from elsewhere), 100% of the Krakataus' biota would be shared with Sebesi but perhaps not all Sebesi's species would be present also on the Krakataus. If there were a significant stepping stone role, therefore, one might expect the proportion of shared species in the Krakataus' biota to be greater than the proportion in Sebesi's biota. Yet this bias would occur even if stepping stone and target had received their biotas from the source independently of one another, with no stepping stone assistance for the target. Being further from the source, the Krakataus may have received only those colonists that are very good dispersers, whereas Sebesi, being nearer the source, would be likely to have these as well as some less efficient dispersers. On its own, therefore, a greater proportion of shared species in the target island would be no proof of a stepping stone role.

If the proportion of shared species were greater in Sebesi's biota than in the Krakataus, this might indicate the importance of Sebesi's stepping stone role, since the proportion of its species that has not colonized the Krakataus would then be less than the proportion of the Krakataus' biota that could not have arrived via the stepping stone. A successful stepping stone should increase the chances of poor dispersers reaching the target, thus increasing the shared proportion on the stepping stone and bringing the proportion of shared species on stepping stone and target closer to one another than they would have been otherwise.

A bias of shared species in favour of Sebesi, however, may also result from the different degrees of survey intensity to which the two areas have been subjected. The Krakataus have received considerably more survey attention than has Sebesi, from both botanists and specialists in various groups of animals, including birds (see above). A 'survey bias' in this direction would tend to increase the shared proportion of Sebesi's biota and decrease the shared proportion of the Krakataus' biota.

Nevertheless the closer together the shared proportions on stepping stone and target, and the greater the stepping stone's shared proportion compared with that of the target, the more effective the stepping stone must have been, and it is of interest to examine these differences for various groups in the Krakatau context.

In seed plants, land molluscs, butterflies, reptiles and bats, the shared pro-portion of Sebesi's fauna is clearly greater than that on the Krakataus, which is substantially lower than 100%. This indicates that Sebesi has been providing some stepping stone role and/or a number of species are colonizing the Krakataus from outside the system without first colonizing Sebesi.

In the case of birds, however, the shared proportions on Sebesi are only very slightly higher (if single Krakatau records are counted) or are rather lower (if they are not) than the shared proportion on the Krakataus. This suggests that Sebesi has been less effective as a stepping stone for land birds than for other groups and/or (less likely) the Krakataus have received a smaller proportion of their avifauna than of other biotic groups from non-Sebesi sources.

Again it is interesting to compare the Sebesi and Panaitan situations. The percentage of shared species in the Panaitan avifauna (52–50%) is lower than that in the Sebesi avifauna (73–69%), which may suggest a more effective step-ping stone role for Sebesi than for Panaitan. This difference may be explained without assuming stepping stone function, however. The poorer dispersers on Panaitan are less likely to reach the Krakataus some 45 km away than are those on Sebesi, only 15 km from the target. Thus the comparison of shared propor-tions on Panaitan and Sebesi provides no evidence for Sebesi's stepping stone role one way or the other. On both stepping stone and non-stepping-stone models Panaitan would be expected to have a lower percentage of species in common with the Krakataus than does Sebesi.

The proportion of Panaitan's avifauna of 66 species that is shared with the Krakataus is 52% (or 50%) and that of the Krakataus' known avifauna of 47 species shared with Panaitan is 64% (or 70%). Again, the pattern for birds differs from those for other groups in the Sebesi context; the bias in shared species is in favour of the target (Krakataus) rather than the Panaitan stepping stone, sug-gesting a small stepping stone role for Panaitan. Moreover, the bias is more marked than is the case in the Sebesi avifaunal comparison (Table 15.6), indicat-ing a greater stepping stone role for Sebesi than for Panaitan as far as the avifauna is concerned. The Sebesi situation of course involves a shorter distance from stepping stone to target and although the stepping stone is further from the mainland there is a second intercalary low and heavily disturbed stepping stone island – Sebuku.

That the proportion of Krakatau's species shared with Sebesi (70–74%) is only slightly higher than that shared with Panaitan (64–70%) could be taken as another indication that Sebesi's stepping stone role has been the greater. Yet this difference may also relate to relative degree of disturbance of the two putative stepping stones and their respective sources. Although Sebesi is in a much more favourable position than Panaitan for the role of stepping stone, it may not have had the overriding influence on the build-up of Krakataus' avifauna that its size and position would suggest.

Asynchronous environmental change on Sebesi and Krakatau

In the context of this theme, the important word in the passage quoted at the beginning of this discussion is 'habitable'. Thus, 'dispersal across gaps of more than a few kilometres is by stepping stones wherever *habitable* stepping stones of even the smallest size exist'. It can be argued that in this case the stepping stone's environment has diverged from that of the target island, making it increasingly uninhabitable for those species likely to colonize the target.

Sebesi's single volcano was forested before 1883 (although only two plants were recorded from the island, *Alangum hexapetalum* and *Serianthus grandiflora*, neither of which has been recorded since). In 1877 there were four coastal villages, *sawahs* (irrigated rice fields), coconuts and pepper gardens (Verbeek 1885, Docters van Leeuwen 1924). The tsunamis associated with the 1883 eruption wreaked destruction over Sebesi's coastal plain and some 2000 people were drowned. After only 2 months (in October 1883), Verbeek (1885) found 'a singularly mournful scene, a picture of utter desolation. The forest has entirely vanished, the thickness of the layers of ashes and pumice stone is from 1–1 1/2 m; everywhere in the ashes deep narrow gullies had been washed out, which ... conveyed an impression as if the mountains were strewn with dead timber. In the plain an occasional tree-stump rears itself above the ashes.' Nothing remained of the four kampongs, 'all has been washed away, and is covered with a layer of ashes 1 metre thick. ... there is no trace left of the tidal wave, as everything is thickly covered with ashes, which fell after the wave. At Seboekoe the height [of the wave] amounts from 25 to 30 metres, but no measurement was taken.' (Verbeek 1884: 13, 14). Cotteau (1885) reported a layer of dried mud 20 feet (6 m) thick on Sebesi in May 1884, and continued (my translation), 'In place of the forest there is only a chaos of whitened trunks lying beneath the hill slopes, partly buried under the ash.' However, although severe, the effects were not as great as they were on the Krakataus themselves. Docters van Leeuwen (1924, 1936) gives depths of the ash layers on Sebuku and Sebesi as 1.5 m, certainly not more than 15 feet (5 m) and on Rakata from about 100 to 200 feet (30–60 m). Nine months after the eruption, Cotteau (1885) noted that on Sebesi plants of *Musa* were growing through the ash from rootstocks and coconuts fallen from pre-existing trees were germinating, whereas there was no sign of plants on the Krakataus. Because the effects of the eruption were less and because it is closer to the reservoir of potential colonizers on Sumatra, Sebesi's recovery was more rapid than that of the Krakataus. Docters van Leeuwen (1923) found that the vegetation differed from that of the Krakataus in lacking extensive grasslands and a distinct *Cyrtandra* zone, and in having a more diverse forest of large trees with lianes. He concluded (p. 162) that in 1921 the vegetation was 'much further developed than on Krakatau'.

The presence of permanent streams on Sebesi and Sebuku has led to human occupation and the resulting disturbance. On Sebesi this began only 7 years after 1883 and has been considerable. Dammerman (1948) and Docters van Leeuwen (1923) both quote a Mr Fontein, a Civil Service Officer at Kalianda, who wrote that in about 1900 15 cattle were introduced to Sebesi and later 20 goats and five horses. About 10 years later some of these escaped and at the time of the Dutch biologists' visits in the early 1920s there were quite large feral populations. Siebers shot a female goat at a height of 700 m in 1921 and Dammerman stated that there were then about 1000 wild cattle on the island. Several villages were present, *ladangs* (non-irrigated rice fields) were formed in the mountains by burning the forest, and fruit plantations were established. Plantations of coconuts, cloves and coffee now cover most of the island.

Sebesi's environment, therefore, began to differ from that on the Krakataus immediately after the 1883 eruption and the two islands' subsequent development has diverged. Sebesi's vegetation changed quickly, recovering from the effects of the 1883 eruption faster than did the Krakataus. The recovery on Sebesi was partly hidden by substantial human-induced changes which began less than a decade after the eruption but did not involve the complication of renewed volcanism since 1930, which has affected recovery on the Krakataus. Sebesi's recovery thus quickly moved 'out of step' with that on the Krakataus. Consequently the ability of Sebesi to support organisms that could colonize the Krakataus will have been gradually declining. Sebesi would be able to act successfully as a stepping stone for fewer and fewer species as the changing environments of stepping stone and target moved increasingly out of synchrony and followed divergent paths. As mentioned above, for this reason it is unlikely that Sebesi was able to make a significant contribution to Krakatau's grassland community. Moreover, because of the increasing agricultural activity and reduction of forest, any stepping stone role Sebesi might have had for forest species would have been diminishing for the past century, whereas the island might have progressively facilitated colonization by ruderal species of more open habitats. However, as Sebesi's role was changing towards that of a potential conduit for open-country species, paradoxically this habitat was declining on the Krakataus. As indicated above and by Runciman *et al.* (1998), most of the Sebesi species that have been unable to colonize the Krakataus appear to have had establishment rather than distributional constraints to their colonization.

Conclusion

There is no evidence on criteria 1 or 2 that Sebesi has had any stepping stone role. Evidence for a preponderance of species of Sumatran origin on the Krakataus, as would be expected if there has been an important stepping stone role, is lacking at present, and there is very little evidence from distributional data that there has been movement of species over time from Sebesi to the Krakataus. On criteria 3 and 4 the distributional data of groups which have been sufficiently well surveyed appear to falsify the hypothesis that Sebesi has had no significant general role as a stepping stone in the colonization of the Krakataus. Although the percentage of its species that is found also on the Krakataus indicates that Sebesi has not acted as a stepping stone for almost half its seed plant and land mollusc species and almost a third of its land birds, it may have done so for the majority of its reptile (80%), butterfly (85%) and bat species (91%). The proportion of species common to stepping stone and target is clearly higher on Sebesi than on the Krakataus except in the case of birds, where the shared proportions on stepping stone and target are about the same. I conclude that Sebesi may have played a significant stepping stone role in the case of seed plants, butterflies, reptiles and bats but it seems likely that this role has been rather less important in the case of land birds.

Although intuitively one would expect Sebesi to have provided an ideal stepping stone for the Krakataus' colonization, the evidence indicates that the importance of the island's stepping stone role has varied across the biota. I suggest that its stepping stone function for almost all groups must have changed over the period of Krakatau's colonization because of the differing and divergent histories of Sebesi and the Krakataus since 1883.

CHAPTER SIXTEEN

Learning from nature's lessons

Community development

Studies of oceanic islands and archipelagos that are millions of years old, such as the Galapagos and Hawaii, can often provide evidence about the dispersal sieves, and sometimes the establishment constraints, that operated in the build-up of island biotas from a zero base (e.g. Thornton 1971, Becker 1975, 1992, Carlquist 1996, Peck 1996). The only good comparison between oceanic and continental islands of similar size in the same region has been by Case and Cody (1987) in the Sea of Cortez. This study was particularly illuminating. They found no difference between continental and oceanic islands in the species richness of land plants, land birds and shore fish, but that land mammals and reptiles were richer on the continental islands. They interpreted this to mean that the two latter groups had not yet declined to equilibrium numbers on the continental islands. Could it also be that these relatively poor dispersers have not yet risen to equilibrium on the oceanic ones?

The relative simplicity of Anak Krakatau's developing community has permitted investigations of the role of particular species in the process of community assembly. For example, Gross (1993) has elucidated the dependence of the pioneer leguminous creeper, *Canavalia rosea*, on its carpenter bee pollinator, *Xylocopa confusa*, and Turner (1992) has investigated the mode of colonization of the island by the antlion *Myrmeleon frontalis*, concluding that human activity assisted not the arrival, but the establishment process. Studies have also been made on the invertebrates associated with another important pioneer plant, *Casuarina equisetifolia* (by Turner 1997).

The colonization of the island by a group of plants that is of prime importance in forest enrichment in the lowland tropics (Lambert and Marshall 1991) – *Ficus* species – has been monitored on Anak Krakatau (Partomihardjo *et al.* 1992) and correlated with the colonization of fruit bats and frugivorous birds (Tidemann *et al.* 1990, Thornton 1992a, Zann and Darjono 1992, Schedvin *et al.* 1995) and agaonine fig wasps (Compton *et al.* 1988, 1994, Thornton *et al.* 1988, 1996).

Recent studies on frugivory in a Bornean lowland rainforest (Shanahan and Compton 2001) confirmed the dietary importance of *Ficus* species to a wide range of frugivores, but they also confirmed that not all species are equal in their suitability for particular frugivores. Throughout the year, some trees of the lowland forest *Ficus* community are in fruit; in monoecious species ripe figs are found year round – this is necessary in order to maintain the supply of pollinating wasps (see below). This 'continuous' phenology contrasts with that of non-*Ficus* fruit-bearing forest trees, which have a 'discontinuous' type of phenology. Thus the *Ficus* community has the potential to provide a resource (which, incidentally, is rich in calcium) for frugivores even through times of low fruit availability, and for this reason has been described as a 'keystone' resource, at least in South-East Asian and Neotropical forests. This, of course, reduces competition between *Ficus* species and members of other fruit-bearing tree genera. Moreover, within the *Ficus* community, different guilds of *Ficus* species attract different subsets of the total frugivore community, thus reducing competition for dispersers between *Ficus* species.

The *Ficus* dispersal guild structure that Shanahan and Compton revealed appeared to be based on interaction between fruit colour, size and height in the forest, with aspects of dispersers' ecology and behaviour. Birds, for example, having good colour vision, are attracted to fig 'fruit' that are red or orange at maturity, carried in the outer canopy and conspicuous against green foliage, whilst fruit bats are attracted to green fruit that often give out a characteristic odour and are often borne close to the trunk in the subcanopy. Based on patterns of frugivore visitation, they identified 'ground mammal', 'fruit bat', 'bulbul', 'primate', 'mixed bird' and 'pigeon' guilds of figs. These tended to be, respectively, borne on or near the ground; green, odorous figs of the understorey; orange–red figs borne in leaf axils in the understorey; large, tough red figs of the subcanopy; small orange–red figs of the canopy; and large orange–red figs of the canopy. The 'ground mammal' and 'primate' guilds are not important in the context of island colonization, for their dispersers do not easily reach islands. The two canopy bird guilds were distinguished mainly on fig size, and the 'pigeon' guild was almost a subset of the 'mixed bird' guild, the former having mostly larger figs, which were taken, in the main, by larger birds that were more effective dispersers. The existence of these guilds is ecologically important, for it ensures that figs whose dispersers seldom if ever reach islands, are not 'wastefully' dispersed to them. Shanahan and Compton noted that before they become propagule packages, the fig syconia are inflorescences, and the necessity that figs attract pollinators may be an important complicating factor in any analysis of *Ficus* fruit characteristics.

Because of their obligate mutualistic and, with very few exceptions, species-specific association with species of small agaonine pollinating wasps, which can survive outside fig syconia for only a few days, the establishment of *Ficus* species on islands appears to have inherent problems. The plant cannot set seed without the intervention of its 'own' species of wasp, which in the vast majority of cases serves no other fig species, and the wasp can only reproduce within the fig 'fruit' of its own species of *Ficus*. Thus, even if a number of seeds of a newly colonizing *Ficus* species have reached an island and germinated and grown to maturity there, that *Ficus* population is nevertheless doomed to sterility unless its specific pollinator reaches the island when figs are at the receptive stage for wasps to enter the syconia to oviposit, and in the process pollinate the flowers. There is now indirect evidence from several sources, including work on Anak Krakatau (Compton *et al.* 1988, 1994), that agaonine wasps are capable of aerial dispersal over distances of scores of kilometres (Thornton *et al.* 1996).

Anak Krakatau offers an ideal testing ground for a theoretical scenario of *Ficus* colonization developed by Bronstein *et al.* (1990) and Bronstein and Patel (1992). They proposed that a newly colonizing *Ficus* species, albeit fruiting in the presence of its specific pollinator, is unlikely to be capable of maintaining its own pollinator population. Adult female wasps emerging from their natal figs must change figs. They must find another one at a younger stage, receptive to wasp entry and oviposition. In a small pioneer colonizing population their chances of succeeding will be very low. This is because in monoecious *Ficus* species, at least, fruiting is usually synchronous within the crown of a tree (preventing selfing) but asynchronous between trees, so there are unlikely to be receptive figs on the wasps' natal tree. Emerging wasps must therefore also change trees. At the low *Ficus* populations prevailing in the first stages of colonization, trees bearing figs that are just at the correct, receptive stage may not happen to be available. At this time, therefore, a fairly continuous immigration of wasps to the island is essential for the maintenance of the mutualism there.

The likelihood of emerging wasps finding receptive figs will increase with the number of trees. At a critical size of the *Ficus* species' population, at least some such receptive figs will always be available to emerging wasps, and the population of *Ficus* species can then maintain its own wasp population. Only then will the fig–wasp mutualism, and therefore the *Ficus* species, have become established on the island.

Preliminary studies of pollination rates of both rare and common pioneer fig species on Anak Krakatau tend to support the theory, and the present episode of volcanic activity, by decimating *Ficus* and fig-wasp populations, may provide a further test (Compton *et al.* 1994, Thornton *et al.* 1996).

Circumstantial evidence raises the possibility that diversification of Anak Krakatau's forest by the immigration of fig species may have been retarded by the colonization, in 1989, of the peregrine falcon, an aerial raptor that specializes on prey species (columbid birds and fruit bats) that constitute the prime dispersers of the most important trees and shrubs of mixed forest (Thornton 1994, Thornton *et al.* 1996).

A sequence of interdependent species may thus form a functional unit of colonization: for example, the animal dispersers of a plant, the plant, its pollinators and their specific predators or parasites, the plant's dependent herbivores and their specific predators and parasites, the plant's specific plant parasites and their dependent animals. Any or all of these may constitute a community depending on the keystone plant for its foundation. *Ficus*, and the major forest tree species on each island, may thus 'guide' the structural development of associated communities which, inevitably, will differ in details of species composition but be unified by key mutualistic relationships. The concept of such 'syncolonization' is part successional and part food chain, but restricted to neither of these.

Rakata and Anak Krakatau

Anak Krakatau's colonization since 1953 has provided, at the same site, a natural colonization model within a larger one (colonization of the archipelago). We therefore compared the course of colonization of Anak Krakatau from 1953 to 1983 with that of Rakata from 1883 to 1923 (Thornton *et al.* 1992). Comparisons were made of colonization by seed plants and pteridophytes (using the (then) latest published flora: Whittaker *et al.* 1989), fig species, resident land birds, frugivorous birds, reptiles, bats (Whittaker and Jones 1994), spiders, braconid hymenopterans and butterflies. The changing dispersal-mode spectra of colonizing plants and the parasitoid mode of braconids were also compared in the two colonizing systems. Species that were not present on the Krakataus for the whole of the period 1953–83, and were thus not available as potential colonizers of Anak Krakatau for the entire three decades, were omitted from the analyses. It was found that earlier, rather than later, colonizers of Rakata between 1883 and 1923 tended to be the successful colonizers of Anak Krakatau since 1953. Moreover, the changes in dispersal-mode spectra of colonizing plants and parasitoid mode of braconids were similar in the two models (Thornton *et al.* 1992).

The general similarities in the colonization process between the two models are significantly greater than would be expected by chance, and indicate a substantial deterministic component of the process, at least up to the time limit of the study, the beginning of mixed forest formation (see the next section). The similarity between the two cases may be explained to some degree by common general climatic conditions and the source biota in one being a selected subset of that in the other. The basis of the early importance of determinism, however, is more likely to be the generally vastly superior powers of dispersal and establishment of pioneer species in the conditions obtaining during this early colonizing phase.

Anak Krakatau and Surtsey

Surtsey and Anak Krakatau are remarkably similar in area (about $2.5\,km^2$), height (some 200 m) and morphology. Both have extensive lava shields which have ensured their permanence, as well as ash-covered areas that include highly mobile lowland promontories subjected to rapid marine erosion and deposition.

Both have neighbouring islands carrying biotas that are subsets of their main-lands' biotas. The closest of the Westman Islands to Surtsey, Geirfuglasker, is 5 km distant, and the largest, Heimaey, is some 20 km away. Anak Krakatau is surrounded by its older pre-1883 companion islands, at a distance of 2–4 km.

There are important differences, however, between Anak Krakatau and Surtsey. Surtsey's colonization probably dates from the end of its phreatic, explosive phase, which lasted for almost 6 months. The vent then became sealed from the sea and two effusive phases followed, with lava flows in 1964–65 and 1966–67. There has been no further volcanic eruption since June 1967. Anak Krakatau's present biota dates from the devastating eruptions in 1952–53, some 22 years after its emergence and, in contrast to Surtsey, its volcanic activity has continued fairly regularly, imposing severe checks on the biota at least three times since then. In spite of the relatively undisturbed conditions for the last 27 years, however, Surtsey's colonization has been much the slower, reflecting the differences in source biota (Iceland has an impoverished biota, having been wiped clean by Pleistocene glaciation) and much harsher physical conditions (Fridriksson and Magnússon 1992).

One group of organisms is so widespread, colonizing such a wide range of environments, that comparison of the faunas of Surtsey and Anak Krakatau is possible – the free-living terrestrial nematodes. Of three bacterial feeders found on Surtsey in 1971, 5 years after the island's emergence, one was *Acrobeloides nanus*, a cosmopolitan, parthenogenetic, unselective feeder able to survive des-iccation, and the other two were species of *Monohystera*. A year later, *Plectus rhizophilus*, a cosmopolitan species known to be extremely resistant to changes in temperature and water availability, was present, and nematodes of the genus *Ditylenchus* were found close to the shore in soil from under a piece of wood covered in fungus. On Rakata, bacterial concentrations in the soil had reached the norm for Java by 1906, 23 years after the eruption (Ernst 1908), and four desiccation-resistant free-living nematodes, including a species of *Plectus*, were found in moss on the island in 1921 (Heinis 1928). By 1984 77 genera were known from the Krakatau archipelago, including all the genera found on Surtsey. The 19 genera present on Anak Krakatau included four genera of the bacterial feeding Cephalobidae (one of which was *Acrobeloides*, found on Surtsey after 8 years) and four Tylenchida, including the fungivorous *Ditylenchus*, repre-sented on Surtsey after 9 years. Both these genera were found on all islands. The unaided dispersal of nematodes is notoriously slow, but the considerable and rather similar faunas found on Anak Krakatau and Surtsey testifies to the ubiquity both of nematodes and suitable carriers.

Repetitions of the colonization process

The repeated destruction and reassembly of Anak Krakatau's biota has provided sequential 'trials' in which conditions were very similar, as good a set of tests of the operation of determinism in the early colonization process as one is likely to find in natural conditions. Of the total of 37 plant species present in two extirpated floras monitored in 1930–34 and 1949–51, 31 (84%) have since recolonized. These are part of a somewhat larger deterministic core of species that is present on all islands of the group and largely consists of sea-dispersed plants of the strand line and coast and a few other wind-dispersed pioneers. In 1992, Anak Krakatau's flora of 140 vascular species included 58 of the 63 sea-dispersed plant species growing on the archipelago, showing that the young island had already received almost the complete complement of this core component of the archipelago's flora (Partomihardjo *et al.* 1992). The assumption that these floras on Anak Krakatau were derived from immigrant propagules has been thrown into question recently by the findings of Whittaker *et al.* (1995) on the viability of buried seeds. From soil samples taken on Panjang and Sertung at depths of 106–175 cm, corresponding to burial by Anak Krakatau ash some 60 years ago, 11 species of seed plants germinated under appropriate nursery conditions. Clearly seeds can remain viable in soil banks for decades following burial in volcanic ash to these relatively shallow depths.

Conclusions from comparisons, including other colonizations

Is it possible to draw general conclusions about the island colonization process by comparing the case histories of the 1883 Krakatau islands, Anak Krakatau and Surtsey? How do the processes occurring on them differ from those known on the few other emergent islands and the barren volcanic substrates that have been studied? In spite of differences in climate, environment, and in the intensity, frequency, and scale of monitoring, some basic similarities emerge.

(1) Cyanobacteria were evident very early in the colonization sequence.
     They were found on Surtsey (Henriksson and Rodgers 1978), on Rakata (Treub 1888), and on bare coral cays studied by Heatwole (1971) in the Coral Sea. On Surtsey at least, and probably also on Rakata, these were nitrogen-fixing and played an important role in the establishment of the pioneer autotrophic plant cover (mosses and ferns, respectively) on nitrogen-deficient substrates.
(2) The initial fauna subsisted on extra-island energy sources that were made available directly, through airborne fallout or through flotsam.
     On cays on Australia's Great Barrier Reef (Heatwole 1971), on Motmot (Bassot and Ball 1972, Ball and Glucksman 1975, 1981, Diamond 1977), on Surtsey (Lindroth *et al.* 1973, Fridriksson 1975, 1994, Bödvarsson 1982) and on Anak Krakatau, the pioneer faunas were dependent on floating detritus. These detritivores and scavengers included collembolans (Anak Krakatau, Surtsey, the bare cays), earwigs (the bare sand cays, Motmot), lycosid spiders (the cays, Motmot) and beetles (bare cays).

On Anak Krakatau another guild of animals, including lycosid spiders, earwigs, a mantis, and a nemobiine cricket, now exploits the airborne fallout of arthropods on barren ash-lava areas and has probably done so since early in the island's history. The *Hierodula* mantis, a member of this guild on Anak Krakatau, was present as part of the island's short-lived fourth fauna. Crickets, earwigs, opilionids, centipedes, reduviid bugs, lizards and a wide variety of spiders comprise this guild on lava fields of the Canary Islands (Ashmole and Ashmole 1988), and nemobiine crickets, earwigs, mantids and lycosid spiders on barren lava on the island of Hawaii (Howarth 1979). There is a hint that earwigs and lycosid spiders filled this role also on Motmot (Ball and Glucksman 1975), and on Mount Etna's a'a lava pholcid and salticid spiders comprise this guild (Wurmli 1974).

(3) Energy from extra-island ecosystems also reaches islands indirectly, through the agency of what Heatwole (1971) called 'transfer organisms'; that is, species acting in the early stages of community assembly as important conduits for the flow of energy from other ecosystems, via sea or air, to the island, by depositing on it waste food, carcasses and faeces.

Seabirds were important in this energy transfer on Surtsey and on the bare sand cays, and on Anak Krakatau green turtles, crabs, monitor lizards and shorebirds filled this role. The monitor was the island's earliest terrestrial vertebrate colonist and the first birds were two resident and six migrant species of shorebirds, five of which (and the sea eagle) were also present only 6 weeks after the devastating 1952–53 eruptions ended (Hoogerwerf 1953b).

On Anak Krakatau dragonflies, swallows, swifts and swiftlets act as transfer organisms for airborne energy by taking invertebrate fallout before it reaches the ground. On the Canary Islands pipits fill this role (Ashmole and Ashmole 1988).

(4) An early-arriving, early-equilibrating, widespread and relatively unchanging plant 'core' comprised a suite of thalassochorous, mainly coastal and strandline plants, and some wind-borne pioneers. Many of the seaborne group are fruit-bearing and are secondarily spread some way inland by vertebrate frugivores.

On Rakata, Anak Krakatau and Surtsey, and on Volcano Island within the crater of Taal volcano in the Philippines (Gates 1914, Brown *et al.* 1917), sea- and wind-dispersed pioneer plant species tended to arrive before zoochorous ones, although on Volcano Island the interval was very short.

(5) As the suite of frugivores exploiting the pioneer plant colonists increased in number and diversified, the zoochorous component of plant colonizers increased.

This is the last plant-dispersal category to equilibrate on Krakatau (it has not yet done so) and includes the vast majority of the woody plants of interior forests, including fig species that are large trees and of great importance in the formation of mixed forest (Thornton 1994, Whittaker and Jones 1994, Thornton *et al.* 1996).

The high proportion of zoochorous plant species in Surtsey's very young flora may be ascribed largely to the establishment of seabird colonies on the island after only 6 years, a situation that has no parallel on the Krakataus. Motmot appears to be exceptional in that its flora, at the time of the last survey, evidently consisted entirely of zoochorous plants; black duck were nesting on the island after only 1 year.

(6)   Closure of the forest canopy affects habitat significantly, altering micro-climate by changing such factors as wind, temperature, relative humidity, desiccation and shade. To survive this major successional change, plant and animal colonists must be adapted to these more mesic conditions. A new, different subset of the source biota is tapped, and those species of the declining open habitat that are unable to adapt to the changed conditions become extinct.

Although on the Krakataus this process was buffered by physical changes to the islands themselves (see above), species that were lost at the time of forest closure include several birds of open habitats such as the peaceful dove, long-tailed shrike and large-billed crow, the gecko *Lepidodactylus lugubris*, sphecid wasps, sapromyzid flies, fulgorid and jassid homopterans and six species of skippers (Lepidoptera: Hesperiidae) that feed as larvae on palms and grasses. Most of the early pioneer heliophilous plants survived, but with a changed role in the system – they became exploitative fillers of forest gaps.

(7)   Approach to an equilibrium condition is far from uniform across compo-nents of the developing biota. This feature has so far been discovered only on the Krakataus as the most intensively investigated case in which such conditions have been sought.

Chance and determinism

The cases briefly reviewed above constitute large-scale 'natural experiments' forming the bases of field studies in community and, in some cases, ecosystem assembly. A number of valuable modelling and laboratory experimental studies have examined aspects of community assembly (e.g. Dickerson and Robinson 1985, Gilpin *et al.* 1986, Robinson and Dickerson 1987, Robinson and Edgemon 1988, Roughgarden 1989, Drake 1990a, 1990b, 1991). How do the laboratory and theoretical findings, and those of field experiments, relate to the processes observed in the natural assembly of biotas?

To recapitulate, in experimental and modelling studies, important factors found to affect the final outcome of the assembly process were colonization rate (reflecting isolation), colonization sequence (priority), length of the interval between species arrivals (colonization interval) and degree of disturbance. In certain circumstances (see below), the last three of these will be highly stochastic. Modellers and laboratory experimentalists found that, using the same pool of colonists, changing these parameters resulted in a limited number of alternative stable communities (with different species and/or dominant relationships). Disturbance, of such a nature and intensity as to affect the context of the colonization process, sometimes led to quite novel species combinations.

There are indications that these factors may operate in nature. For example, the laboratory studies confirmed one of Maguire's (1963) findings from simple field experiments on the colonization of small artificial water bodies, that only a limited number of species combinations were able to form stable communities. On the Krakataus, also, there are some indications that some of the factors found to have importance from laboratory and modelling studies may have affected the course and outcome of colonization significantly. The development of different forest types (dominated by different tree species) on Rakata, Sertung and Panjang has been ascribed to the different sequences of arrival and inter-arrival intervals of a few species of potentially dominant forest trees (Tagawa *et al.* 1985), and to Anak Krakatau's continuing volcanic activity differentially deflecting successional paths on these islands (Whittaker *et al.* 1989, 1992a). It is likely that both these factors have been involved (Tagawa 1992, Thornton 1995). Moreover, the substantial disturbance caused by Anak Krakatau's volcanic activity after the late 1920s may have been partly responsible for the assembly, on Rakata, Sertung and Panjang, of forest communities that have no counterparts on the mainlands, or indeed elsewhere, reflecting the conclusions of Drake (1990a), for example.

The deterministic force that is perhaps most frequently cited in studies of island colonization is a biological one – the competitive exclusion of one species by interacting species. Competition was included in the equilibrium theory of island biogeography (MacArthur and Wilson 1967; see also Williamson 1982), and competitive exclusion hierarchies and networks formed the basis of Diamond's assembly rules (Diamond 1975). Competitive exclusion is recognized almost always on the basis of indirect evidence – usually the presence of only one of two or more species that are available in the source pool, are capable of dispersing to the island, and are known or are assumed to have requirements for the same limiting resources. Possible examples on the Krakataus are few (Thornton 1996b), suggesting that this factor operates largely in the later stages of colonization, perhaps after the attainment of a dynamic equilibrium.

Since 1979 there has been considerable discussion on the relative roles of deterministic and stochastic processes in community assembly (e.g. Connor and Simberloff 1979, Gilpin and Diamond 1982, Lawton 1987, Wissel and Maier 1992). Case and Cody's (1987) study of the biotas of continental and oceanic islands in the Sea of Cortez was a rare instance of an investigation of natural situations throwing some light on this question, although the light only revealed that generalizations about the development of island biotas are almost impossible to make, with each island situation having unique features that control its own peculiar colonization process. In some cases competition appears to have played a major role in determining the resulting biota, in others it has been unimportant and habitat has been the main determinant. Our comparison of the colonization of Anak Krakatau with that of Rakata, outlined in an earlier section, also showed that the colonization process up to forest formation had a considerable deterministic component, in the sense that in a population of 'Anak Krakatau islands' the communities assembled would be expected to differ little between islands. Conditions for establishment were evidently more important than the random dispersal of the various species in determining the island's resulting biota.

Heatwole and Levins (1973) conducted an exciting study in the late 1960s which not only provided interesting (and rare) data on dispersal as well as turnover rates, but also gave insights into the early stages of community formation. Between 1964 and 1966 they made 12 surveys of a small vegetated sand cay, Cayo Ahogado, lying 1 km off the coast of Puerto Rico. Heatwole and Levins estimated that at one time or another during this period nine plant species and 18 species of animals were established on the cay.

Heatwole and Levins documented a high turnover rate for established plants (1.6 species per year) and an extinction rate of 0.8 species per year. As a minimum, an average of 14.4 species per year were newly arrived on the cay but the plant immigration rate falls to 1.6 species per year if only successfully colonizing species (i.e. those that became established) are counted. Establishment was defined as reproductive success on the cay and presence over at least two consecutive survey periods. Heatwole and Levins concluded that extinction was due to lack of suitable habitat, wave erosion and hurricanes. The equilibrium species number of established plants was about seven species. Only five species remained in the flora throughout the study, and all are sea-dispersed: *Sesuvium portulacastrum*, *Cakile lanceolatum*, *Ipomoea pes-caprae*, *Atriplex tetranda* and *Vigna repens*. Two species, *Thespesia populnea* (which is known to be sea-dispersed) and *Cyperus planifolia*, became extinct but recolonized, and two, *Panicum repens* and *Euphorbia buxifolia*, colonized during the study period.

During the period of the surveys nine lizard individuals arrived, representing three species. The arrival times of lizard individuals correlated with flood periods on Puerto Rico, when the frequency of natural rafts floating out to sea was at its highest. With one possible exception, male and female of the same species did not happen to be on the island at the same time, and none of the species became established. The cay was visited after heavy rains and associated frequent flotsam in 1970 by McKenzie *et al.* (1971), when there were 13 lizards, of the same three species found previously, but now two of the species (*Anolis cristatellus* and *A. pulchellus*) were represented by both sexes and juveniles and the third (*Ameiva exsul*) by an adult and juveniles. Some (or all) of these species may have become established.

A total of 77 species of terrestrial invertebrates, one marine water strider and three semiterrestrial crustaceans were recorded from Cayo Ahogado, and the number of established species of animals fluctuated around an equilibrium of about ten species. An isopod (*Philoscia* species), an earthworm and a chernetid pseudoscorpion were probably present throughout the study period. Three spider species were thought to have become established at one time or another during the study (*Tennesseelum formicum*, *Steatoda septemmaculata* and *Lycosa atlantica*), and five ant species. Seven other species were present in more than one survey and were regarded as having become established: two tenebrionid beetle species (one a species of *Phaleria*), a staphylinid beetle, a labidurid earwig (*Anisolabis maritima*), a lygaeid bug (*Exptochiomera minima*) which had also been found alive in in flotsam, a cricket (a species of *Cycloptilum*) and a psocopteran (*Ectopsocus vilhenai*), a member of a group that exploits ephemeral fungal food sources on temporary habitats such as dead leaves. All the other invertebrates were recorded only once except for two ants and a strongly flying pierid butterfly (*Appias* species) which was seen repeatedly at sea.

The species of animals and plants that became established were almost all transported on flotsam or floated themselves, and Heatwole and Levins (1972a) showed that flotsam was an important dispersal mechanism for many invertebrates on the Puerto Rico Bank. Even the four spider species recorded from the cay are all ground spiders found beneath objects. Heatwole and Levins thought it unlikely that they had arrived on the air, and more probable that they had come to the cay in flotsam. Transport of propagules by wind and by human activity were thought to be relatively unimportant in the colonization of the cay.

Heatwole and Levins' (1972b) analysis of the Florida mangrove experiments of Simberloff and Wilson (1969, 1970) showed that when the defaunated mangrove cays were recolonized, the trophic spectrum or structure (number of feeding hierarchies and relative numbers of species within them) that obtained before defaunation tended to be quickly re-established, although the original taxonomic structure is not achieved until after a much longer period (Simberloff and Wilson 1970). A particular trophic structure thus appeared to be a substantially deterministic, stable feature of a community developed on a given ecological substrate. If this is so, then trophic structure on an island such as Cayo Ahogado, where turnover of species is high, should be more stable than taxonomic structure. To test this, Heatwole and Levins determined the trophic structure of the cay's fauna at five survey periods during the first year of the study, and compared changes with changes in taxonomic structure. Trophic structure clearly varied less than did taxonomic structure.

A second prediction from the concept of a deterministic, stable trophic structure is that a putative colonist's establishment on an island will depend on the immigrant species fitting into the trophic web, either by occupying a vacant food niche or by outcompeting and displacing a species already present. In turn this will mean that the trophic structure of the island community will be a non-random sample of the trophic spectrum of the putative colonizing species arriving. As is always the case, Heatwole and Levins had insufficient data on species that arrived at the cay but failed to become established, but they did have data on animals found alive on flotsam from the area of the Puerto Rican Bank, many from near Cayo Ahogado (Heatwole and Levins 1972a). In spite of the fact that the trophic level of many of these animals was not determinable, there was one striking difference between the two. There was a high frequency of wood-boring animals in the flotsam, yet wood-borers were completely absent from Cayo Ahogado. Thus the cay's fauna was not a random sample of the possible colonists in the flotsam.

Heatwole and Levins also compared the mean trophic structure of the Cayo Ahogado community over the 12-month period with the structure of the non-vegetated sand cays in the Coral Sea that Heatwole (1971) had studied, and with the original structure of Florida mangrove cays that Simberloff and Wilson (1969) had used in their defaunation studies. The mangrove cays were similar to one another in trophic structure, as were the non-vegetated sand cays, and trophic structure of Cayo Ahogado was stable over time. As might have been expected, the mangrove cays had a higher proportion of herbivores than did the vegetated sand cay Cayo Ahogado, and the unvegetated Coral Sea cays had the lowest proportion. The reverse trend was seen for scavengers and detritus feeders. These were important on the sand cays, less so on Cayo Ahogado, and relatively unimportant on the mangrove cays. Each of the three cay categories therefore had a characteristic trophic structure which was related to the relative availability of energy from plants and dead organic matter. It seems therefore that on small cays the establishment of immigrants and the number of species present are determined in part by the type of trophic structure which the island's available energy resources can support.

Heatwole (1981) also found that the trophic structure of the fauna of One Tree Island in Australia's Great Barrier Reef remained remarkably stable in the face of great change in taxonomic composition over a period of 3 years. He concluded that interspecific interactions such as competition and predation control the community's trophic structure (that is, the proportions of species in various trophic guilds) but chance seems to determine which species make up that structure. Moreover, Heatwole found that islands with the same type of food base (for example, the marine environment or terrestrial plants) had the same trophic structure. The number of different kinds of food resources seemed to determine the proportions of species in the various feeding guilds, and the abundance of the food determines the number of individuals within the trophic categories.

Heatwole (1981) and Heatwole and Levins (1972b, 1973) thus showed that constraints on trophic structure may also provide a deterministic element in the colonization process. A deterministic aspect of community assembly associated with trophic levels was also hinted at by Maguire (1963) (see above).

Both stochastic and deterministic processes may have to be invoked to explain observed variation in trophic patterns in small isolated freshwater communities – those inhabiting plant-held waters (phytotelmata: Kitching 2000) – each relating to a different scale of observation. At the regional scale, deterministic factors (general environmental predictability or energetic limitations at the base of the food web) appear to control the trophic structure but at the local scale stochastic factors best explain the observed variation.

From volcanic islands to water-filled tree holes may seem like a giant leap, but island biogeographers need to keep a wide perspective! Phytotelmata are relatively simple, natural islands, available in multiplicate. Such natural island microcosms are study units that have much to offer the student of island communities. Moreover, modelling approaches such as that of Post and Pimm (1983), for example, have almost equal relevance to community ecologists studying water-filled tree holes and to island biogeographers. In seeking to understand the processes we study on 'real islands' we need to remain aware of, and where appropriate make connection with, the work of theoreticians and of laboratory and field community ecologists. What is needed is not merely vicariance, but some strong, intellectually robust, long-distance dispersal between these fields.

In order to identify stochastic components in the competitive exclusion situation Seamus Ward and I constructed a simple model involving two interacting species, assuming that the outcome of the interaction (in terms of the persistence of one or the other species on the island) depends only on sequence of colonization (priority) and the interval between colonizations of the species (colonizing interval). We assumed that there was a critical period required on the island by one species, free of the other, after which the other could not become established (the critical interval). The critical interval will depend on the relative competitive abilities of the two species, and on species attributes such as reproductive rate that influence initial establishment in the absence of the interacting species.

Our model showed that both priority and the relation between colonizing interval and the critical interval may contribute to stochasticity, and that contributions from these sources vary over the period of colonization. Stochasticity due to variation in priority is low when species have similar, high arrival rates (as do early pioneer species) or when they differ greatly in arrival rate (for example pioneer species compared with later-successional species), but high

Table 16.1 *Number of vascular plant species in the two successions on Motmot and Anak Krakatau, and the number and percentage of species in the first succession (S1) that were present in the second succession (S2)*

	S1	S2	In common
Motmot	22	44	9 (41%)
Anak Krakatau	43	170	37 (86%)

where species have similar, low arrival rates (as do many later-colonizing animal-dispersed trees, for example). Where species differ considerably but not extremely in dispersal ability (as do many species in the later stages of colonization), then, although a given species may be likely to arrive before another and priority-based stochasticity will be low, colonizing interval will vary, and the species' critical interval may or may not fall within it. Stochasticity resulting from variation in colonizing interval may be high in the late stages of colonization. These results support the findings from comparative studies that the early course of colonization is highly deterministic, and it appears likely that stochasticity will increase during the stage of forest formation, decline again during the second phase of wind-dispersed immigration at the time of canopy closure, and then rise once more until a dynamically stable state is reached.

The repeated destruction and reassembly of Anak Krakatau's biota provided sequential 'trials' in which conditions were very similar, as good a set of tests of the operation of determinism in the early colonization process as one is likely to find in natural conditions. Of the total of 37 plant species present in two extirpated floras monitored in 1930–34 and 1949–51, 31 (84%) have since recolonized. These are part of a somewhat larger deterministic core of species, present on all the islands of the group, largely consisting of sea-dispersed plants of the strandline and coast and a few other wind-dispersed pioneers. At this time Anak Krakatau's flora of 140 vascular species included 58 of the 63 sea-dispersed plant species growing on the archipelago, showing that the young island had already received almost the complete complement of this core component of the archipelago's flora (Partomihardjo *et al.* 1993) (Table 16.1).

Anak Krakatau's colonization since 1953, many aspects of which have been discussed in this book, has provided a natural colonization model within a larger one (colonization of the archipelago). Colonization of Anak Krakatau from 1953 to 1983 (the last published flora, Whittaker *et al.* 1989) was compared with colonization of Rakata from 1883 to 1923 (Thornton *et al.* 1992). Species that were not present on the Krakataus for the whole of this period, and thus had limited availability as potential colonizers of Anak Krakatau, were omitted from the analysis. Comparisons were made of colonization by seed plants, pteridophytes, fig species, resident land birds, frugivorous birds, reptiles, bats, spiders, braconid hymenopterans and butterflies. The changing dispersal-mode spectra of colonizing plants and the parasitoid mode of braconids were also compared in the two colonizing systems. There was a significant tendency for earlier, rather than later, colonizers of Rakata between 1883 and 1923 to be the successful colonizers of Anak Krakatau since 1953, and the change in dispersal-mode spectra of colonizing plants and parasitoid mode of braconids were similar in the two models.

Thus there were more similarities in the colonization process generally between the two models than would be expected by chance, indicating a significant deterministic component, at least up to the time limit of the study, the beginning of mixed forest formation. The similarity between the two can be explained to some degree by common general climatic conditions and the source biota in one being a selected subset of that in the other. The basis of the early importance of determinism (in the sense that in a population of 'Anak Krakatau islands' the communities assembled would be expected to differ little between islands), however, is more likely to be the generally vastly superior powers of dispersal and establishment of pioneer species in the conditions obtaining during this early phase.

In most situations the process of community assembly is likely to involve both stochastic and deterministic elements. Early in the process, when arrival rates of pioneer species are similar and high (and much different from those of non-pioneers) and colonizing intervals are short, stochasticity will be low, in the sense that in a population of similar islands the communities that would be assembled would be similar to one another.

If species have fairly similar, low arrival rates (e.g. pairs of species on distant islands), priority will vary a great deal and the outcome of colonization will be highly dependent on it. In this situation stochasticity will be high. On the three older Krakatau islands the course of development of mixed forest diverged after the 1920s. On Panjang and Sertung, but not on Rakata, each some 4 km from one another, areas of forest were dominated by one of two animal-dispersed trees, *Timonius compressicaulis* and *Dysoxylum gaudichaudianum*. Although here dominance, rather than exclusion, is the effective outcome, the pattern appears to support conclusions from the model. On Rakata the colonizing interval between these species was at least 34 years and on Panjang at most 3 years, illustrating the highly stochastic nature of colonizing interval when two relatively poor dispersers are involved.

Where the species differ considerably, but not extremely, in dispersal ability, priority would be expected to be important but to vary little between cases and thus to contribute little to stochasticity. Our model showed that variation in colonizing interval, rather than variation in priority, may now have contributed to stochasticity. In the Rakata forests the wind-dispersed tree *Neonauclea calycina*, a species with a dispersal rate higher than the two zoochorous species

by perhaps two orders of magnitude, became dominant. On Rakata the colonizing intervals between *Neonauclea* and *Timonius* and between *Neonauclea* and
Dysoxylum were at least 24 and 34 years respectively, but on Panjang the corresponding intervals were at least zero and 3 years, and on this island *Timonius* and
Dysoxylum became the dominants. The observed outcome is not at variance with
the model's prediction that variable colonizing interval, rather than varying
priority, provided stochasticity in this case.

Thus because in the forest-forming stage of succession there is likely to be
more variability both in priority and in colonizing interval, stochasticity is likely
to be higher than in the early pioneer phase.

As forest formation proceeded, however, and the canopy closed, a second
wave of aerially dispersed species colonized the islands. Presumably its component species were determined largely by high dispersal powers and the strict
demands placed on species by the new, more mesic environment of the forested
island, and colonization returned to a more deterministic mode.

The last colonizing wave comprised largely zoochorous trees and the positive
feedback associated with this dispersal mode led to a steady rise in its importance. In this wave immigration rates are relatively low and similar for all
potential colonists, and as seen on the Krakataus in the case of *Dysoxylum* and
Timonius, both priority and colonizing interval provide high stochasticity, resulting in variation between cases.

Both theoretical (Drake 1990a) and experimental (Robinson and Dickerson
1987, Robinson and Edgemon 1988) studies of community assembly have shown
that species' probabilities of successful colonization may depend in part on
their sequence of colonization. As MacArthur (1972) recognized, this means
that there are families of immigration and extinction curves rather than single
curves. It is surprising that both experimental and comparative biogeographical
data (species area curves) show that although colonization sequence may
change the form of the curves, it does not appear to affect the final equilibrium
species number. While equilibrium species richness in model communities may
depend on colonization sequence (Drake 1990a), this has not been borne out in
experimental studies. Robinson and Dickerson (1987) found that colonization
sequence had a significant but small (8%) effect on species richness only at high
invasion rates, and in the studies of Robinson and Edgemon (1988), species
richness was far more sensitive to overall invasion rate and extent of the interval
between species' arrivals than to the sequence of invasions.

In the experimental and modelling studies cited above, degree of disturbance
was found to be one important factor affecting the final outcome of the assembly process, along with colonization rate (i.e. isolation), colonization sequence
(priority) and length of the interval between species arrivals (colonization interval). Using the same pool of colonists, modellers and laboratory experimentalists both found that changing these parameters (including disturbance) resulted

in a limited number of alternative stable communities (that is, with different species and/or dominant relationships). Disturbance of such a nature and intensity as to affect the context of the colonization process sometimes led to quite novel species combinations.

There are indications that these factors may operate in nature, where (see below) the degree of disturbance may itself be largely a matter of chance. On the Krakataus, also, there are some indications that some of the factors found important in laboratory and modelling studies may have affected the course and outcome of colonization significantly. We have seen that the development of different forest types (dominated by different tree species) on Rakata, Sertung and Panjang has been ascribed to the different sequences of arrival and inter-arrival intervals of the few species of potentially dominant forest trees (Tagawa *et al.* 1985), but it has also been ascribed to Anak Krakatau's continuing volcanic activity differentially deflecting successional paths on these islands (Whittaker *et al.* 1989, 1992a). It is likely that both these factors have been involved (Tagawa 1992, Thornton 1996b). Moreover, the substantial disturbance caused by Anak Krakatau's volcanic activity after the late 1920s may have been partly responsible for the assembly, on Rakata, Sertung and Panjang, of forest communities that have no counterparts on the mainlands, or indeed elsewhere.

It has been suggested (Bush and Whittaker 1993: 456) that intermediate- to high-magnitude disturbance events such as hurricanes 'may cast the system away from equilibrium, such that it is "permanently" non-equilibrial'. In those systems with alternative equilibria, any occasional gross disturbance large enough to cause mass extinction would drive species richness from a higher equilibrium into the domain of attraction of a lower one, but it seems that few natural disturbances – and very few hurricanes – are of sufficient magnitude to do this.

On the Krakataus different guilds, functional groups and taxa, which appear to be approaching equilibrium at different rates (Thornton *et al.* 1990, 1993, Thornton 1991), also appear to be differentially susceptible to disturbances. Preliminary observations on the effects of the 1992–94 episode of volcanic activity of Anak Krakatau (Thornton *et al.* 1994) indicated that components of the island community are differentially susceptible to frequent falls of volcanic ash. Analyses of the effects of Caribbean and Central American hurricanes (e.g. Brokaw and Walker 1991, Waide 1991) also revealed differential susceptibility of community components to disturbance. These analyses, moreover, provided no evidence that species richness was substantially affected as a result of the hurricanes. Likewise, in Nicaraguan primary rainforest, 4 months after a 1988 hurricane which destroyed the canopy over a vast area, there was little or no change in species composition in spite of this drastic alteration of the forest structure. Recovery involved the direct regeneration of primary forest species rather than a return to an early phase of succession (Yih *et al.* 1991).

For a number of reasons, the obligate mutualistic association between figs and their species-specific pollinating wasps may be expected to be highly susceptible to population crashes of either mutualist. Bronstein and Hossaert-McKey (1995) have recently shown, however, that even this highly specialized and seemingly fragile interaction can be very resilient to population-level catastrophies. In Florida in August 1992, Hurricane Andrew, one of the most severe hurricanes of the century, stripped census fig trees (*Ficus aurea*) of all leaves, figs and pollinating wasps. Within 5 months the flowering phenology of the fig had returned to pre-hurricane patterns, the pollinator had recolonized, probably from an unaffected source some 60 km away, and the restoration of asynchronous flowering had enabled it to achieve pre-hurricane population levels.

In practice, therefore, the only event likely to have the effects that Bush and Whittaker ascribe to a hypothetical Hurricane Joe is a cataclysmic eruptive event of Krakatau 1883 scale and intensity, in which case, however, the system would return to zero or near zero and the process would begin again.

Limited recruitment may have an important effect on competition – making it so unimportant that even competitively inferior species may colonize in spite of their inherently superior competitors, because the inferior species are superior in the early colonizing stages. Thus even when the biota consists of highly competitive component species, limitation of recruitment may reintroduce and strengthen the element of chance.

Summary of the colonization process

The following stages have been observed in a number of situations, although in no single case have they all been evident. The timing is approximate.

(1) First few years. Separate guilds of pioneer scavenging, predacious animals colonize, exploiting seaborne organic detritus and wind-borne arthropods. Stochasticity low.

(2) Plant pioneers. Seaborne shore and strand plants form a coastal woodland by two decades, which equilibrates by three. Wind-borne heliophilous herbaceous plants such as orchids, ferns and grasses form a grassland by two to three decades, also early equilibration. Stochasticity low.

(3) Euryphagous, widely distributed, open-country animals colonize in the first few decades.

(4) The first generalist avian frugivores, arriving within the first 25 years, and fruit bats, within four decades, exploit some of the shore plants in (2), leading to their spread further inland. Other frugivores bring pioneer heliophilous fruiting trees, which together with those of (2) form inland patches of trees – open woodland. Stochasticity high.

(5) Animal colonists now include specialized frugivorous birds, predators and insectivorous bats. Further animal-dispersed trees and shrubs arrive,

shading out the grassland; forest is formed and its canopy closes after about 50 years. Stochasticity high.

(6)   A second wave of wind-dispersed herbaceous plants comprises mesophilous, shade-tolerant, species including many ferns and orchids. Pioneer plant species persist as gap-fillers. Turnover of animal species reaches a peak as shade-adapted mesophilous species colonize and heliophilous open-country species are lost. Some herbivores dependent on grassland plants are lost. Stochasticity low.

(7)   Addition of further animal-dispersed large-seeded trees, and very few wind-dispersed species, with winged seeds, is slow. Predators arrive. True forest animals and plants colonize. Stochasticity high. Equilibration of animal groups begins with those with good dispersal and without specific constraints on establishment.

Light spots

Towards the end of his masterful work *The Diversity of Life*, E. O. Wilson (1992) reviewed the importance and value of Earth's biodiversity, and the present threats to what he described as our most valuable but least appreciated resource. He then listed several enterprises that should be put into place in order to save and use in perpetuity as much of that resource as possible. These comprised the following strategies:

(1)   Survey the world's flora and fauna.
(2)   Create biological wealth. (By appreciating and utilizing the economic, medical, agricultural, recreational and aesthetic value of our natural biodiversity.)
(3)   Promote sustainable development. (On a global scale, through national decisions on, and implementation of, an optimal human population in the context of the world's population, economy and environment.)
(4)   Save what remains. (By increasing the 'reserved' percentage of the land surface and including as many undisturbed habitats as possible, including identified 'hots pots'.)
(5)   Restore the wildlands. (By regenerating existing ecosystems that are endangered.)

The ongoing, long-term study of the origins and development of island ecosystems, what may be termed 'developmental ecology', is pertinent to Wilson's strategies 4 and 5. The young emergent islands treated in this book (Surtsey, Anak Krakatau, Motmot, Tuluman), together with the islands recovering from recent devastation (Volcano Island, Bárcena, Krakataus, Long Island, Ritter), demonstrate how nature goes about the process of ecosystem assembly from a zero base-line and ecosystem restoration following complete or near-complete devastation.

First, they need to be protected (strategy 4), so that the development process can proceed without interference. Surtsey and the Kameni islands are already well served in this regard by their governments; Surtsey has been cherished by Iceland from its inception (and recently (2006) has been nominated, long over-due, for listing as a World Heritage Site) and Santorini has volcanological and archaeological interest, as well as tourist appeal. The other national govern-ments involved (Indonesia, Papua New Guinea, Philippines, Mexico) are doing all they can, but in order to carry out their responsibilities properly they need top-level international encouragement and support, on a scale at least equiv-alent to that provided for Ecuador's curatorship of the Galapagos, for example.

Second, good funding should be provided to qualified researchers so that the studies can continue. All the emergent islands and most of the devastated, recovering ones need to be monitored at maximum intervals of 5 years or so. These field studies should be properly funded, and viewed as long-term invest-ments in understanding ecological processes. There is a tendency for granting bodies to think that such field studies can be undertaken successfully on shoe-string budgets. They are right, of course. Clearly successful expeditions can be achieved with very limited resources. But provision of fast, safe transport, and use-as-needed funding for emergencies (such as provision for helicopter use) not only save time, but allow scientists to get on with their job free from unnecessary logistic and safety worries. Here is a striking example. One of our Krakatau team visited the islands with a television film company a few years after our latest expedition. He enjoyed the luxury of a fast, safe, modern vessel plying daily from Java across the 50-km stretch of Sunda Strait to the Krakataus in about an hour, for 2 weeks. On the expeditions that helped to provide the scientific basis for the film, we had to use 'selected' fishing vessels. The safest we could find were slow and suffered chronic mechanical failure. Several of our expeditioners still tell hair-raising stories of journeys (lasting, in storm condi-tions up to 8 hours or so) on these boats, together with problems of transporting all our needs for 2–4-week stays on the islands. If money can be provided for the human entertainment aspect, it should surely also be made available for the globally significant scientific aspect. And it needs to be remembered that the work does not stop when fieldwork ends – indeed, one could argue that that is when the work really starts. Funding should take this into account, but very rarely does so. Perhaps scientific cooperation with the entertainment industry should be conditional upon grants from the industry for future field research!

Norman Myers coined the term 'hot spots' in 1988 for areas of the world that contain large numbers of endemic species and are under threat. He listed 18 examples that satisfied this definition. The concept has proved popular in help-ing to designate countries for which conservation need is both great and likely to be relatively cost-effective in terms of 'numbers of species per dollar'. Later, Myers *et al.* (2000) increased this list to 25 regions. Only three of them are entire islands – Madagascar, Sri Lanka and New Caledonia – although designated archipelagos such as the Philippines, Sundaland, Wallacea and Polynesia/Micronesia each comprise numerous others. The small emergent and recovering islands considered in this book would not satisfy the definition of hot spots as individual entities, because they neither contain large numbers of endemic species (they are too young) nor great biological diversity. Moreover, by and large, they are not under imminent extreme threat nor, to my knowledge, do any of them carry globally endangered species. Yet they contain a heuristic resource that is extremely valuable in helping to interpret how the natural world functions.

Perhaps a second global list should be drawn up to include areas or islands that warrant protection because, for 'accidental', historical reasons, they have become natural laboratories for the study of ecological processes that we may well need to understand in the future, in order to ensure ecosystem sustainability. The list would presumably include islands such as Hawaii, Galapagos, Juan Fernandez, New Caledonia and Madagascar, which are natural laboratories for the study of evolution (and thus New Caledonia and Madagascar would be on both lists), but would surely include also the emergent and recovering islands that feature in this book. Since inclusion in the second list would not require qualification as 'hot spots', perhaps they could be designated 'light spots', for the enlightenment that they may bring to us in pursuing Wilson's fifth strategy – 'Restore the wildlands'.

In Central America, tropical dry forest is even more threatened than rainforest; it is now down to only 2% of its original extent. Daniel Janzen's group in Costa Rica have attempted something that would have daunted even the most ardent 'Greenie'. In sparsely populated farmland they have established a 50 000-hectare area, the newly-created Guanacaste National Park, in which they intend to reassemble the dry tropical forest ecosystem. They are using remnant forest patches as seeding centres from which to regenerate forest over the intervening ranchland (Janzen 2000, 2004).

But what if matters have deteriorated to such a degree that there are not even small remnant reservoirs from which to expand? Would it be possible to re-create an ecosystem from scratch? This may have to be attempted in future, perhaps, eventually, on a large scale. Continuing studies of the ways that this has been achieved naturally in the very few 'light spots' of the world may surely be enlightening.

References

Abe, T. (1984). Colonization of the Krakatau Islands by termites (Insecta: Isoptera). *Physiological Ecology, Japan* **21**, 63–88.

Abe, T. (1987). Evolution of life types in termites. In Kawano, S., Connell, J. H. and Hidaka, T. (eds.) *Evolution and Coadaptation in Biotic Communities*. Tokyo, University of Tokyo Press, pp. 124–148.

Andersen, D. C. and MacMahon, J. A. (1985). The effects of catastrophic ecosystem disturbance: the residual mammals at Mount St Helens. *Journal of Mammalogy* **66**, 581–589.

Anderson, M. A. (1978/9). Comments on the presence of a crocodile or crocodiles in Lake Wisdom, on Long Island north of New Guinea. *Science in New Guinea* **6**, 6–8.

Anderson, N. H. (1992). Influence of disturbance on insect communities in Pacific Northwest streams. *Hydrobiologia* **248**, 79–92.

Anonymous (1990). General discussion. In Hardy, D. A. and Renfrew, A. C. (eds.) *Thera and the Aegean World III*, vol. 3, *Chronology*. London, The Thera Foundation, pp. 236–241.

Antos, J. A. and Zobel, D. B. (1987). How plants survive burial: a review and initial responses to tephra from Mount St. Helens. In Bilderback, D. E. (ed.) *Mount St. Helens 1980: Botanical Consequences of the Explosive Eruption*. Berkeley, CA, University of California Press, pp. 246–261.

Ashmole, M. J. and Ashmole, N. P. (1988). Arthropod communities supported by biological fallout on recent lava flows in the Canary Islands. *Entomologica Scandanavica* **332** (Suppl.), 67–88.

Ashmole, N. P. and Ashmole, M. J. (1997). The land fauna of Ascension Island: new data from caves and lava flows, and a reconstruction of the prehistoric ecosystem. *Journal of Biogeography* **24**, 549–589.

Ashmole, N. P., Ashmole, M. J. and Oromi, P. (1990). Arthropods of recent lava flows on Lanzarote. *Vieraea* **18**, 171–187.

Ashmole, N. P., Oromi, P. and Ashmole, M. J. (1992). Primary faunal succession in volcanic terrain: lava and cave studies on the Canary Islands. *Biological Journal of the Linnean Society* **46**, 207–234.

Asong, J. (1984). How tuna fish came to be: a legend from the Siassi Islands, Morobe Province. *Paradise Magazine* (in-flight with Air Niugini) **47** (July 1984), 17.

Aston, M. A. and Hardy, P. G. (1990). The pre-Minoan landscape of Thera: a preliminary statement. In Hardy, D. A., Keller, J., Galanopoulos, V. P., Flemming, N. C. and Druitt, T. H. (eds.) *Thera and the Aegean World III*, vol. 2, *Earth Sciences*. London, The Thera Foundation, pp. 348–361.

Backer, C. A. (1929). *The Problem of Krakatoa as Seen by a Botanist*. Surabaya, Indonesia, published by author.

Baillie, M. G. L. (1990). Irish tree rings and an event in 1628 BC. In Hardy, D. A. and Renfrew, A. C. (eds.) *Thera and the Aegean World III*, vol. 3, *Chronology*. London, The Thera Foundation, pp. 160–166.

Baillie, M. G. L. and Munro, M. A. R. (1988). Irish tree rings, Santorini and volcanic dust veils. *Nature* **332**, 344–346.

Ball, E. (1977). Life among the ashes. *Australian Natural History* **19**, 12–17.

Ball, E. E. (1982a). Long Island, Papua New Guinea: European exploration and recorded contacts to the end of the Pacific War. *Records of the Australian Museum* **34**, 447–461.

Ball, E. E. (1982b). Annotated bibliography of references relating to Long Island, Papua New Guinea. *Records of the Australian Museum* **34**, 527–547.

Ball, E. E. and Glucksman, J. (1975). Biological colonization of Motmot, a recently created tropical island. *Proceedings of the Royal Society of London B* **190**, 421–442.

Ball, E. E. and Glucksman, J. (1978). Limnological studies of Lake Wisdom, a large New Guinea caldera lake with a simple fauna. *Freshwater Biology* **8**, 455–468.

Ball, E. E. and Glucksman, J. (1980). A limnological survey of Lake Dakataua, a large caldera lake on West New Britain, Papua New Guinea, with comparisons to Lake Wisdom, a younger nearby caldera lake. *Freshwater Biology* **10**, 73–84.

Ball, E. E. and Glucksman, J. (1981). Biological colonization of a newly created volcanic island and limnological studies on New Guinea lakes, 1972–1978. *National Geographic Society Research Reports* **13**, 89–97.

Ball, E. E. and Hughes, I. M. (1982). Long Island, Papua New Guinea: people, resources and culture. *Records of the Australian Museum* **34**, 463–525.

Ball, E. E. and Johnson, R. W. (1976). Volcanic history of Long Island, Papua New Guinea. In Johnson, R. W. (ed.) *Volcanism in Australasia*. Amsterdam, Elsevier, pp. 133–147.

Bassot, J. M. and Ball, E. E. (1972). Biological colonization of a recently created island in Lake Wisdom, Long Island, Papua New Guinea, with observations on the fauna of the lake. *Proceedings of the Papua New Guinea Science Society* **23**, 26–35.

Beard, S. J. (1976). The progress of plant succession on the La Soufrière of St Vincent: observations in 1972. *Vegetatio* **31**, 69–77.

Beaver, R. A. (1977). Non-equilibrium 'island' communities: Diptera breeding in dead snails. *Journal of Animal Ecology* **46**, 783–798.

Becker, P. (1975). Island colonization by carnivorous and herbivorous Coleoptera. *Journal of Animal Ecology* **44**, 893–906.

Becker, P. (1992). Colonization of islands by carnivorous and herbivorous Heteroptera and Coleoptera: effects of island area, plant species richness, and 'extinction' rates. *Journal of Biogeography* **19**, 163–171.

Beehler, B. M., Pratt, T. K. and Zimmerman, D. A. (1986). *Birds of New Guinea*. Princeton, NJ, Princeton University Press.

Blong, R. J. (1975). The Krakatoa myth and the New Guinea highlands. *Journal of the Polynesia Society* **84**, 213–217.

Blong, R. J. (1979). Huli legends and volcanic eruptions, Papua New Guinea. *Search* **10**, 93–94.

Blong, R. J. (1982). *The Time of Darkness*. Canberra, Australian National University Press.

Blot, C. (1978). Volcanism and seismicity in Mediterranean island arcs. In *Thera and the Aegean World I*, Papers presented at the 2nd International Scientific Congress, Antorini, Greece, August 1978. London, Thera and the Aegean World, pp. 33–44.

Bödvarsson, H. (1982). The Collembola of Surtsey, Iceland. *Surtsey Research Progress Reports* **9**, 63–67.

Bonaccorso, F. (1998). *Bats of Papua New Guinea*. Washington DC, Conservation International.

Bonaccorso, F. J. and McNab, B. K. (1997). Plasticity of energetics in blossom bat (Pteropodidae): impact on distribution. *Journal of Mammalogy* **78**, 1073–1088.

Brattstrom, B. H. (1963). Barcena Volcano, 1952: its effect on the fauna and flora of San Benedicto Island, Mexico. In Gressitt, J. L. (ed.) *Pacific Basin Biogeography*. Honolulu, HI, Bishop Museum Press, pp. 499–524.

Brattstrom, B. H. (1990). Biogeography of the Islas Revillagigedo, Mexico. *Journal of Biogeography* **17**, 177–183.

Bristowe, W. S. (1931). A preliminary note on the spiders of Krakatau. *Proceedings of the Zoological Society of London* **1931**, 1387–1442.

Bristowe, W. S. (1934). Introductory notes. *Proceedings of the Zoological Society of London* **1934**, 11–18.

Brokaw, N. V. L. and Walker, L. R. (1991). Summary of the effects of Caribbean hurricanes on vegetation. *Biotropica* **23**, 442–447.

Bromenshenk, J. J., Postle, R. C., Yamasaki, G. M., Felli, D. G. and Reinhardt, H. E. (1987). The effect of Mount St. Helens ash on the development and mortality of the western spruce budworm and other insects. In Bilderback, D. E. (ed.) *Mount St. Helens 1980: Botanical Consequences of the Explosive Eruption*. Berkeley, CA, University of California Press, pp. 302–317.

Bronstein, J. L. and Hossaert-McKey, M. (1995). Hurricane Andrew and a Florida fig pollination mutualism: resilience of an obligate interaction. *Biotropica* **27**, 373–381.

Bronstein, J. L. and Patel, A. (1992). Causes and consequences of within-tree phenological patterns in the Florida strangling fig, *Ficus aurea* (Moraceae). *American Journal of Botany* **79**, 41–48.

Bronstein, J. L., Guoyon, P. H., Gliddon, C., Kjellberg, F. and Michaloud, G. (1990). The ecological consequences of flowering asynchrony in monoecious figs: a simulation study. *Ecology* **71**, 2145–2156.

Brown, J. H. and Kodric-Brown, A. (1977). Turnover rates in insular biogeography: effect of immigration on extinction. *Ecology* **58**, 445–449.

Brown, W. H., Merrill, E. D. and Yates, H. S. (1917). The revegetation of Volcano Island, Luzon, Philippine Islands, since the eruption of Taal volcano in 1911. *Philippine Journal of Science, Series C, Botany* **12**, 177–248.

Buckland, P. C., Dugmore, A. J. and Edwards, K. J. (1997). Bronze Age myths? Volcanic activity and human response in the Mediterranean and North Atlantic regions. *Antiquity* **71**, 581–593.

Burt, W. H. 1961. Some effects of Volcan Paricutin on vertebrates. *Occasional Papers of the Museum of Zoology of the University of Michigan* **620**, 1–24.

Bush, M. B. (1986). The butterflies of Krakatoa. *Entomologist's Monthly Magazine* **122**, 51–58.

Bush, M. B. and Whittaker, R. J. (1991). Krakatau: colonization patterns and hierarchies. *Journal of Biogeography* **18**, 341–356.

Bush, M. B. and Whittaker, R. J. (1993). Non-equilibration in island theory of Krakatau. *Journal of Biogeography* **20**, 453–457.

Bush, M. B., Bush, D. J. B. and Evans, R. D. (1990). Butterflies of Krakatau and Sebesi: new records and habitat relations. In Whittaker, R. J., Asquith, N. M., Bush, M. B. and Partomihardjo, T. (eds.) *Krakatau Research Reports 1989*. Oxford, School of Geography, University of Oxford, pp. 35–41.

Bush, M. B., Whittaker, R. J. and Partomihardjo, T. (1992). Forest development on Rakata, Panjang and Sertung: contemporary dynamics (1979–1989). *GeoJournal* **28**, 185–199.

Cadogan, G. (1990). Thera's eruption into our understanding of the Minoans. In Hardy, D. A., Doumas, C. G., Sakellarakis, J. A. and Warren, P. M. (eds.) *Thera and the Aegean World III*, vol. 1, *Archaeology*. London, The Thera Foundation, pp. 93–96.

Carlquist, S. (1965). *Island Life: A Natural History of the Islands of the World*. Garden City, NY, The Natural History Press.

Carlquist, S. (1996). Plant dispersal and the origin of Pacific island floras. In Keast, A. and Miller, S. E. (eds.) *The Origin and Evolution of Pacific Island Biotas, New Guinea to Eastern Polynesia: Patterns and Processes*. Amsterdam, SPB Academic Publishers, pp. 153–164.

Case, T. J. and Cody, M. L. (1987). Testing theories of island biogeography. *American Scientist* **75**, 402–411.

Clausen, H. B., Hammer, C. U., Hvidberg, C. S. *et al.* (1997). A comparison of the volcanic record over the past 4000 years from the Greenland Ice Core Project and Dye3 Greenland ice cores. *Journal of Geophysical Research* **102**, 26707–26723.

Cogger, H. (2000). *Reptiles and Amphibians of Australia*, 5th edn. Sydney, Reed Books.

Compton, S. G., Thornton, I. W. B., New, T. R. and Underhill, L. (1988). The colonization of the Krakatau Islands by fig wasps and other chalcids (Hymenoptera, Chalcidoidea). *Philosophical Transactions of the Royal Society of London B* **322**, 459–470.

Compton, S. G., Ross, S. and Thornton, I. W. B. (1994). Pollinator limitation of fig tree reproduction on the island of Anak Krakatau (Indonesia). *Biotropica* **26**, 180–186.

Connor, E. F. and Simberloff, D. (1979). The assembly of species communities: chance or competition? *Ecology* **60**, 1132–1140.

Cook, S., Singidan, R. and Thornton, I. W. B. (2001). Colonization of an island volcano, Long Island, Papua New Guinea, and an emergent island, Motmot, in its caldera lake. IV. Colonization by non-avian vertebrates. *Journal of Biogeography* **28**, 1353–1363.

Cooke, R. J. S. (1981). Eruptive history of the volcano at Ritter Island. *Geological Survey of Papua New Guinea Memoir* **10**, 115–123.

Cooke R. J. S., McKee, C. O., Dent, V. F. and Wallace, D. A. (1976). A striking sequence of volcanic eruptions in the Bismarck volcanic arc, Papua New Guinea, in 1972–75. In R. W. Johnson (ed.) *Volcanism in Australasia*. Amsterdam, Elsevier, pp. 149–172.

Corbet, G. B. and Hill, J. E. (1991). *A World List of Mammalian Species*. Oxford, Oxford University Press.

Cotteau, E. (1885). Borneo and Krakatoa II. A week at Krakatoa. *Proceedings of the New South Wales and Victoria Branches of the Royal Geographical Society of Australia* **2**, 103–106.

Coultas, N. F. (1935). Unpublished journal and letters of William F. Coultas, Whitney South Sea Expedition, vol. IV, October 1933–March 1935. In Department of Ornithology, American Museum of Natural History, New York.

Crawford, R. L., Sugg, P. M. and Edwards, J. S. (1995). Spider arrival and primary establishment on terrain depopulated by volcanic eruption at Mount St. Helens, Washington. *American Midland Naturalist* **133**, 60–75.

D'Addario, G. W. (1972). The 1968 eruption of Long Island. In Johnson, R. W., Taylor, G. A. M. and Davies, R. A. (eds.) *Geology and Petrology of Quaternary Volcanic Islands off the North Coast of New Guinea*, Bureau of Mineral Resources, *Geology and Geophysics Record no. 21*. Canberra, Commonwealth of Australia, Department of National Development, pp. 110–112.

Dammerman, K. W. (1922). The fauna of Krakatau, Verlaten Island and Sebesy. *Treubia* **3**, 61–112.

Dammerman, K. W. (1948). The Fauna of Krakatau 1883–1933. *Verhandlungen Koninlijk Nederlandsche Akademie Wetenschaft* **44**, 1–594.

de Boer, J. L. and Sanders, D. T. (2002). *Volcanoes in Human History: The Far-Reaching Effects of Major Eruptions.* Princeton, NJ, Princeton University Press.

Decker, R. W. (1990). How often does a Minoan eruption occur? In Hardy, D. A., Keller, J., Galanopoulos, V. P., Flemming, N. C. and Druitt, T. H. (eds.) *Thera and the Aegean World III*, vol. 2, *Earth Sciences*. London, The Thera Foundation, pp. 444–452.

del Moral, R. and Bliss, L. C. (1987). Initial vegetation recovery on subalpine slopes of Mount St. Helens, Washington. In Bilderback, D. E. (ed.) *Mount St. Helens 1980: Botanical Consequences of the Explosive Eruption.* Berkeley, CA, University of California Press, pp. 148–167.

del Moral, R. and Bliss, L. C. (1993). Mechanisms of primary succession: insights resulting from the eruption of Mount St. Helens. *Advances in Ecological Research* **24**, 1–66.

del Moral, R. and Wood, D. M. (1986). Subalpine vegetation recovery five years after the Mount St. Helens eruptions. In Keller, S. A. C. (ed.) *Mount St. Helens: Five Years Later.* Spokane, WA, Eastern Washington University Press, pp. 215–221.

del Moral, R., Titus, J. H. and Cook, A. M. (1995). Early primary succession on Mount St. Helens, Washington, USA. *Journal of Vegetation Science* **6**, 107–123.

Diamond, J. M. (1970). Ecological consequences of island colonization by southwest Pacific birds. *Proceedings of the National Academy of Sciences of the USA* **67**, 529–563.

Diamond, J. M. (1972). Reconstitution of bird community structure on Long Island, New Guinea, after a volcanic explosion. *National Geographical Society Research Reports* **13**, 191–204.

Diamond, J. M. (1973). Distributional ecology of New Guinea birds. *Science* **179**, 759–769.

Diamond, J. M. (1974a). Colonization of exploded volcanic islands by birds: the supertramp strategy. *Science* **184**, 803–806.

Diamond, J. M. (1974b). Recolonization of exploded volcanic islands by New Guinea birds. *Exploration Journal*, March, 2–10.

Diamond, J. M. (1975). Assembly of species communities. In Cody, M. L. and Diamond, J. M. (eds.) *Ecology and Evolution of Communities.* Cambridge, MA, Harvard University Press, pp. 342–444.

Diamond, J. M. (1976). Preliminary results of ornithological exploration of the islands of Vitiaz and Dampier Straits, Papua New Guinea. *Emu* **76**, 1–7.

Diamond, J. M. (1977). Colonization of a volcano inside a volcano. *Nature* **270**, 13–14.

Diamond, J. M. (1981). Reconstitution of bird community structure on Long Island, New Guinea, after a volcanic explosion. *National Geographical Society Research Reports* **1**, 191–204.

Diamond, J. M. (1989). This-fellow frog, name belong-him Dakwo. *Natural History*, April, 16–23.

Diamond, J. M., Bishop, K. D. and van Balen, S. (1987). Bird survival in an isolated Javan woodland: island or mirror? *Conservation Biology* **1**, 132–142.

Diamond, J. M., Pimm, S. L., Gilpin, M. E. and LeCroy, M. (1989). Rapid evolution of character displacement in myzomelid honeyeaters. *American Naturalist* **413**, 675–708.

Diapoulis, C. (1971). The development of the flora of the volcanic islands Palaia Kammeni and Nea Kammeni. *Acta 1st International Congress on Volcano Thera* **1969**, 238–247.

Dickerson, J. E. Jr and Robinson, J. V. (1985). Microcosms as islands: a test of the

MacArthur–Wilson equilibrium theory. *Ecology* **66**, 966–980.

Docters van Leeuwen, W. M. (1923). The vegetation of the island of Sebesy, situated in the Sunda Strait, near the islands of the Krakatau group, in the year 1921. *Annales du Jardin Botanique de Buitenzorg* **32**, 135–192.

Docters van Leeuwen, W. M. (1924). On the present state of the vegetation of the Krakatau Group and of the island of Sebesy. *Proceedings of the 2nd Pan Pacific Science Congress* **1923**, 313–318.

Docters van Leeuwen, W. M. (1936). Krakatau 1883–1933: a botany. *Annales du Jardin Botanique de Buitenzorg* **46–47**, 1–507.

Dommergues, Y. (1966). La fixation symbiotique de l'azote chez les Casuarina. *Annales de l'Institut Pasteur* **111** (Suppl.), 247–259.

Downey, W. and Tarling, D. (1984). The end of the Minoan civilization. *New Scientist*, 13 September, 49–52.

Drake, J. A. (1990a). Communities as assembled structures: do rules govern pattern? *Trends in Ecology and Evolution* **5**, 159–164.

Drake, J. A. (1990b). The mechanics of community assembly and succession. *Journal of Theoretical Biology* **147**, 213–233.

Drake, J. A. (1991). Community-assembly mechanics and the structure of an experimental species ensemble. *American Naturalist* **137**, 1–26.

Druitt, T. H. and Francaviglia, V. (1990). An ancient caldera cliff line at Phira, and its significance for the topography and geology of pre-Minoan Santorini. In Hardy, D. A., Keller, J., Galanopoulos, V. P., Flemming, N. C. and Druitt, T. H. (eds.) *Thera and the Aegean World III*, vol. 2, *Earth Sciences*. London, The Thera Foundation, pp. 362–369.

Dwyer, P. D. (1978). A study of *Rattus exulans* (Peale) (Rodentia: Muridae) in the New Guinea Highlands. *Australian Wildlife Research* **5**, 221–248.

Dwyer, P. D. (1984). From garden to forest: small rodents and plant succession in Papua New Guinea. *Australian Mammalogy* **7**, 29–36.

Edwards, J. S. (1986). Derelicts of dispersal: arthropod fallout on Pacific northwest volcanoes. In W. Danthanarayana (ed.) *Insect Flight: Dispersal and Migration*. Berlin, Springer-Verlag, pp. 196–203.

Edwards, J. S. (1987). Arthropods of alpine aeolian ecosystems. *Annual Review of Entomology* **32**, 163–179.

Edwards, J. S. (1988). Life in the allobiosphere. *Trends in Ecology and Evolution* **3**, 111–114.

Edwards, J. S. (1996). Arthropods as pioneers: recolonization of the blast zone on Mt. St. Helens. *Northwest Environmental Journal* **2**, 63–74.

Edwards, J. S. (2005) Animals and volcanoes: survival and revival. In Martí, J. and Ernst, G. L. (eds.) *Volcanoes and the Environment*. Cambridge, Cambridge University Press, pp. 250–272.

Edwards, J. S. and Banko, P. C. (1976). Arthropod fallout and nutrient transport: a quantitative study of Alaskan snowpatches. *Arctic and Alpine Research* **8**, 237–245.

Edwards, J. S. and Schwarz, L. M. (1981). Mount St. Helens ash: a natural insecticide. *Canadian Journal of Zoology* **59**, 714–715.

Edwards, J. S. and Sugg, P. (1993). Arthropod fallout as a resource in the colonization of Mount St. Helens. *Ecology* **74**, 954–958.

Edwards, J. S. and Thornton, I. W. B. (2001). Colonization of an island volcano, Long Island, Papua New Guinea, and of an emergent island, Motmot, in its caldera lake. VI. A pioneer arthropod community on Motmot. *Journal of Biogeography* **28**, 1379–1388.

Edwards, J. S., Crawford, R. L., Sugg, P. M. and Peterson, M. A. (1986). Arthropod recolonization in the blast zone of Mount St. Helens. In Keller, S. A. C. (ed.) *Mount St. Helens: Five Years Later*. Spokane, WA, Eastern Washington University Press, pp. 329–333.

Eggler, W. A. (1948). Plant communities in the vicinity of the volcano El Paricutin, Mexico, after two and a half years of eruption. *Ecology* **29**, 415–436.

Eggler, W. A. (1959). Manner of invasion of volcanic deposits by plants with further evidence from Paricutin and Jorullo. *Ecological Monographs* **29**, 268–284.

Eggler, W. A. (1963). Plant life of Paricutin volcano, Mexico, eight years after activity ceased. *American Midland Naturalist* **69**(12), 38–68.

Ernst, A. (1908). *The New Flora of the Volcanic Island of Krakatau*. London, Cambridge University Press.

Evans, A. (1921–35). *The Palace of Minos at Knossos*, 4 vols. London, Macmillan.

Evans, G. (1940). The characteristic vegetation of recent volcanic islands in the Pacific. *Bulletin of Miscellaneous Information, Botanic Gardens, Kew* **1939**, 43–44.

Fisher, R. V. and Schminke, H.-U. (1984). *Pyroclastic Rocks*. Berlin, Springer-Verlag.

Fisher, R. V., Heiken, G. and Hulen, J. B. (1997). *Volcanoes: Crucibles of Change*. Princeton, NJ, Princeton University Press.

Flannery, T. F. (1995). *Mammals of New Guinea*. Sydney, Reed Books.

Fosberg, F. R. (1985). Botanical visits to Krakatoa in 1958 and 1963. *Atoll Research Bulletin* **292**, 39–45.

Fouqué, F. (1879). *Santorin et ses Éruptions*. Paris, G. Masson.

Fouqué, F. (1998). *Santorini and its Eruptions*, translation and comments by A. R. McBirney. Baltimore, MD, Johns Hopkins University Press.

Frederiksen, H. B., Pedersen, A. L. and Christensen, S. (2000). Substrate-induced respiration and microbial growth in soil during the primary succession on Surtsey, Iceland. *Surtsey Research Progress Reports* **11**, 29–35.

Freidrich, W. L. (2000). *Fire in the Sea: The Santorini Volcano – Natural History and the Legend of Atlantis*, translation by A. R. McBirney. Cambridge, Cambridge University Press.

Freidrich, W. L., Seidenkrantz, M.-S. and Nielsen, O. B. (2000). Santorini (Greece) before the Minoan eruption: reconstruction of the ring-island natural resources and clay deposits from the Akrotiri Excavation. In McGuire, W. J., Griffiths, D. R., Hancock, P. L. and Stewart, I. (eds.) *The Archaeology of Geological Catastrophes*. London, Geological Society, pp. 71–80.

Fridriksson, S. (1975). *Surtsey: Evolution of Life on a Volcanic Island*. London, Butterworth.

Fridriksson, S. (1987). Plant colonization of a volcanic island: Surtsey, Iceland. *Arctic and Alpine Research* **19**, 425–431.

Fridriksson, S. (1992). Vascular plants on Surtsey 1981–1990. *Surtsey Research Progress Reports* **10**, 17–30.

Fridriksson, S. (1994). *Surtsey: Lifríki í Mótun (Surtsey: the development of life on a young volcanic island)*. Reykjavik, Society of Natural History and Surtsey Research Society. (In Icelandic)

Fridriksson, S. (2000). Vascular plants on Surtsey 1991–98. *Surtsey Research Progress Reports* **11**, 21–28.

Fridriksson, S. and Magnússon, B. (1992). Development of the ecosystem on Surtsey with reference to Anak Krakatau. *GeoJournal* **28**, 287–291.

Frith, H. J. (1982). *Pigeons and Doves of Australia*. Adelaide, Rigby.

Fritsch, F. E. (1931). Some aspects of the ecology of freshwater algae. *Journal of Ecology* **19**, 232–272.

Frör, E. and Beutler, A. (1970). The herpetofauna of the oceanic islands in the Santorini archipelago, Greece. *Spixiana* **1**, 301–308.

Fytikas, M., Kolios, N. and Vougioukalakis, G. (1990). Post-Minoan volcanic activity of the Santorini Volcano: volcanic hazard and risk, forecasting possibilities. In Hardy, D. A., Keller, J., Galanopoulos, V. P., Flemming, N. C. and Druitt, T. H. (eds.) *Thera and the Aegean World III*, vol. 2, *Earth Sciences*. London, The Thera Foundation, pp. 183–198.

Galanopoulos, A. G. (1971). The eastern Mediterranean trilogy in the Bronze Age. *Proceedings of the International Scientific Congress on the Volcano of Thera*, Athens, pp. 184–210.

Galloway, R. B. and Liritzis, Y. (1992). Provenance of Aegean volcanic tephras by high resolution gamma-ray spectroscopy. *Nuclear Geophysics* **6**, 405–411.

Gates, F. C. (1914). The pioneer vegetation of Taal Volcano. *Philippine Journal of Science, Series C, Botany* **9**, 391–434.

Gathorne-Hardy, F. J., Jones, D. T. and Mawdsley, N. A. (2000). The recolonization of the Krakatau islands by termites (Isoptera) and their biogeographical origins. *Biological Journal of the Linnean Society* **71**, 252–267.

Gennardus, Br. (Balvers, L. F.) (1983). Enkele waarnemingen aan verschillenke Kweken van rupien en vlinders na een vulkanische asregen op Java. *Entomologische Berichten* **43**, 69–71.

Gersich, F. M. and Brusven, A. M. (1982). Volcanic ash accumulation and ash-voiding mechanisms of aquatic insects. *Journal of the Kansas Entomological Society* **55**, 290–296.

Gilham, M. E. (1970). Seed dispersal by birds. In Perring, F. (ed.) *The Flora of a Changing Britain*. Faringdon, E. W. Classey, pp. 909–98.

Gilpin, M. E. (1980). The role of stepping-stone islands. *Theoretical Population Biology* **17**, 247–253.

Gilpin, M. E. and Diamond, J. M. (1982). Factors contributing to non-randomness in species co-occurrences on islands. *Oecologia* **52**, 75–84.

Gilpin, M., Carpenter, M. P. and Pomeranz, M. J. (1986). The assembly of a laboratory community: multispecies competition in *Drosophila*. In Diamond, J. M. and Case, T. J. (eds.) *Community Ecology*. New York, Harper and Row, pp. 23–40.

Gjelstrup, P. (2000). Soil mites and collembolans on Surtsey, Iceland, 32 years after the eruption. *Surtsey Research Progress Reports* **11**, 43–50.

Gorschkov, S. G. (ed.) (1974). *Atlas of the Oceans: Pacific Ocean*. Moscow, Ministry of Defence, USSR Navy.

Greuter, W. (1979). The origins and evolution of island floras as exemplified by the Aegean Archipelago. In Bramwell, D. (ed.) *Plants and Islands*. London, Academic Press, pp. 87–93.

Grimm, R. (1981). Die Fauna der Ägäis-Insel Santorin. II. Tenebrionidae (Coleoptera). *Stuttgarter Beiträge zur Naturkunde A, Biologie* **348**, 1–14.

Gross, C. L. (1993). The reproductive ecology of *Canavalia rosea* (Fabaceae) on Anak Krakatau, Indonesia. *Australian Journal of Botany* **41**, 591–599.

Hall, G. (1987). Seed dispersal by birds of prey. *Zimbabwe Science News* **21**, 9.

Halvorson, J. J., Smith, J. L. and Franz, E. H. (1991). Lupine influence on soil carbon, nitrogen and microbial activity in developing ecosystems at Mount St. Helens. *Oecologia* **87**, 162–170.

Hammer, C. U., Clausen, H. B. and Dansgaard, W. (1980). Greenland ice sheet evidence of post-glacial volcanism and its climatic impact. *Nature* **288**, 230–235.

Hammer, C., Clausen, H., Friedrich, W. and Tauber, H. (1987). The Minoan eruption of Santorini in Greece dated to 1645 BC? *Nature* **328**, 517–519.

Hansen, A. (1971). Flora der Inselgruppe Santorin. *Candollea* **26**, 109–163.

Harding, A. F. (1989). Blind dating. *New Scientist*, 6 May, 52–53.

Hardy, D. A. and Renfrew, A. C. (eds.) 1990. *Thera and the Aegean World III*, vol. 3, *Chronology*. London, The Thera Foundation.

Harrison, R. D., Yamuna, R. and Thornton, I. W. B. (2001). Colonization of an island volcano, Long Island, Papua New Guinea, and of an emergent island, Motmot, in its caldera lake. II. The vascular flora. *Journal of Biogeography* **28**, 1131–1137.

Heatwole, H. (1971). Marine-dependent terrestrial biotic communities on some cays in the Coral Sea. *Ecology* **52**, 363–366.

Heatwole, H. (1981). *A Coral Island: The History of One Tree Reef*. Sydney, Collins.

Heatwole, H. and Levins, R. (1972a). Biogeography of the Puerto Rican Bank: flotsam transport of terrestrial animals. *Ecology* **53**, 112–117.

Heatwole, H. and Levins, R. (1972b). Trophic structure stability and faunal change during recolonization. *Ecology* **53**, 531–534.

Heatwole, H. and Levins, R. (1973). Biogeography of the Puerto Rican Bank: species turnover on a small cay, Cayo Ahogado. *Ecology* **54**, 1042–1055.

Heiken, G. and McCoy, F. (1990). Precursory activity to the Minoan eruption, Thera, Greece. In Hardy, D. A., Keller, J., Galanopoulos, V. P., Flemming, N. C. and Druitt, T. H. (eds.) *Thera and the Aegean World III*, vol. 2, *Earth Sciences*. London, The Thera Foundation, pp. 79–88.

Heiken, G., McCoy, F. and Sheridan, M. (1990). Palaeotopographical and palaeogeological reconstruction of Minoan Thera. In Hardy, D. A., Keller, J.,

Galanopoulos, V. P., Flemming, N. C. and Druitt, T. H. (eds.) *Thera and the Aegean World III*, vol. 2, *Earth Sciences*. London, The Thera Foundation, pp. 370–376.

Heinis, F. (1928). Die Moosfauna des Krakatau. *Treubia* **10**, 231–244.

Heldreich, T. von (1899). Die Flora der Insel Thera. In Hiller von Gärtringen, F. (ed.) *Die Insel Thera in Altertum und Gegenwart*, vol. 1. Berlin, Georg Reoimer, pp. 122–140.

Heldreich, T. von (1902). Die Flora der Insel Thera. In Hiller von Gärtringen, F. (ed.) *Die Insel Thera in Altertum und Gegenwart*, vol. 4. Berlin, Georg Reoimer, pp. 119–130.

Hendrix, L. B. and Smith, S. D. (1986). Post-eruption revegetation of Isla Fernandina, Galapagos. II. *National Geographic Research* **2**, 6–16.

Henriksson, L. E. and Rodgers, G. A. (1978). Further studies in the nitrogen cycle of Surtsey, 1974–1976. *Surtsey Research Progress Reports* **8**, 30–40.

Hodkinson, I. D., Coulson, S. J., Webb, N. R. *et al.* (1996). Temperature and the biomass of flying midges (Diptera: Chironomidae) in the high Arctic. *Oikos* **75**, 241–248.

Hodkinson, I. D., Coulson, S. J., Harrison, J. and Webb, N. R. (2001). What a wonderful web they weave: spiders, nutrient capture and early ecosystem development in the high Arctic – some counter-intuitive ideas on community structure. *Oikos* **95**, 349–352.

Hommel, P. W. F. M. (1987). *Landscape Ecology of Ujung Kulon (West Java, Indonesia)*. Wageningen, Soil Survey Institute.

Hoogerwerf, A. (1953a). Some notes about the nature reserve Pulau Panaitan (Prinseneiland) in Strait Sunda with special reference to the avifauna. *Treubia* **21**, 481–505.

Hoogerwerf, A. (1953b). Notes on the vertebrate fauna of the Krakatau Islands, with special reference to the birds. *Treubia* **22**, 319–353.

Housley, R. A., Hedges, R. E. M., Law, I. A. and Bronk, C. R. (1990). Radiocarbon dating by AMS of the destruction of Akrotiri. In Hardy, D. A. and Renfrew, A. C. (eds.) *Thera and the Aegean World III*, vol. 3, *Chronology*. London, The Thera Foundation, pp. 207–215.

Howarth, F. G. (1979). Neogeoaeolian habitats on new lava flows on Hawaii island: an ecosystem supported by windborne debris. *Pacific Insects* **20**, 133–144.

Howarth, F. G. (1987). Evolutionary ecology of aeolian and subterranean habitats in Hawaii. *Trends in Ecology and Evolution* **2**, 220–223.

Howarth, F. G. and Montgomery, S. L. (1980). Notes on the ecology of the high altitude aeolian zone on Mauna Ke'a. *'Elepaio* **41**, 21–22.

Ibkar-Kramadibrata, H., Soeriaatmadja, R. E., Syarif, H. *et al.* (1986). *Explorasi Biologis dan Ecologis dari Daerah Daratan di Gugus Kepulauan Krakatau menjelang 100 tahun sesudah Peletusan*. Bandung, Institut Teknologi.

Jacobson, E. R. (1909). Die nieuwe fauna van Krakatau. *Jaarverslag van der Topographischen Dienst Nederlandsch-Indie* **4**, 192–211.

Janzen, D. H. (2000). Costa Rica's Area de Conservación Guanacaste: a long march to survival through non-damaging biodevelopment. *Biodiversity* **1**, 7–20.

Janzen, D. H. (2004). Setting up tropical biodiversity for conservation through non-damaging use: participation by parataxonomists. *Journal of Applied Ecology* **41**, 181–187.

Jimenez, C., Ortega-Rubio, A., Alvcarez-Cardenas, S. and Arnaud, G. (1994). Ecological aspects of the land crab *Geocarcinus planatus* (Decapoda: Gecarcinidae) in Socorro island, Mexico. *Biological Conservation* **69**, 9–13.

Johnson, R. W. (1976). Late Cainozoic volcanism and plate tectonics at the southern margin of the Bismarck Sea, Papua New Guinea. In R. W. Johnson (ed.) *Volcanism in Australasia*. Amsterdam, Elsevier, pp. 101–116.

Johnson, R. W. and Smith, I. E. (1974). Volcanoes and rocks of St Andrews Strait, Papua New Guinea. *United States Naval Medical Bulletin* **46**, 1628–1632.

Johnson, R. W., Taylor, G. A. M. and Davies, R. A. (1972). *Geology and Petrology of Quaternary Volcanic Islands off the North Coast of New Guinea*, Bureau of Mineral Resources, Geology and Geophysics Record no. 21. Canberra, Commonwealth of Australia, Department of National Development.

Johnstone, B. (1997). Who killed the Minoans? *New Scientist* **154**, 36–39.

Jones, P. (1986). The bryophytes of the Krakatau Islands. *Department of Geography, University of Hull, Miscellaneous Series* **33**, 77–86.

Jones, P. D., Briffa, K. R. and Schweingruber, F. H. (1995). Tree-ring evidence of the widespread effects of explosive volcanic eruptions. *Geophysical Research Letters* **22**, 1333–1336.

Keenan, D. J. (2003). Volcanic ash residue from the GRIP ice core is not from Thera. *Geochemistry, Geophysics, Geosystems* **4**(110) 1097. doi:10.1029/3003GC000608.

Keller, S. A. C. (ed.) (1986). *Mount St. Helens: Five Years Later*. Spokane, WA, Eastern Washington University Press.

Kingsley, C. (1915). *The Water Babies*. London, Constable.

Kisokau, K. (1974). Analysis of avifauna stomach contents of Long and Crown Islands, Madang District. *Science in New Guinea* **2**, 261–262.

Kisokau, K., Pohei, Y. and Lindgren, E. (1984). *Tuluman Island after Thirty Years: An*

Inventory of Plants and Animals of Tuluman Island, Manus Province. Boroko, Office of Environment and Conservation, Papua New Guinea.

Kitching, R. L. (2000). *Food Webs and Container Habitats: The Natural History and Ecology of Phytotelmata*. Cambridge, Cambridge University Press.

Koopman, K. F. (1979). Zoogeography of mammals from islands of the northeastern coast of New Guinea. *American Museum Novitates* **2690**, 1–17.

Krafft, M. (1991). *Volcanoes: Fire from the Earth*. London, Thames and Hudson.

Kristinsson, H. (1974). Lichen colonization in Surtsey, 1971–1973. *Surtsey Research Progress Reports* **7**, 9–16.

Kuniholm, P. I. (1990). Overview and assessment of the evidence for the date of the eruption of Thera. In Hardy, D. A. and Renfrew, A. C. (eds.) *Thera and the Aegean World III*, vol. 3, *Chronology*. London, The Thera Foundaton, pp. 13–18.

Kuniholm, P. I., Kromer, B., Manning, S. W. *et al.* (1996). Anatolian tree rings and the absolute chronology of the eastern Mediterranean, 2220–718 BC. *Nature* **381**, 780–783.

Kunkel, G. (1981). *Die Kanarischen Inseln und ihre Pflanzenwelt*. Stuttgart, Gustav Fischer.

Kuwayama, S. (1929). Eruption of Mt Komagatake and insects. *Kontyu* **3**, 271–273.

Lack, D. (1976). *Island Biology, Illustrated by the Land Birds of Jamaica*. Oxford, Blackwell Scientific Publications.

La Marche, V. C. and Hirschboeck, K. K. (1984). Frost rings in trees as records of major volcanic eruptions. *Nature* **307**, 121–126.

Lamb, H. H. (1970). Volcanic dust in the atmosphere; with its chronology and assessment of its meteorological significance. *Philosophical Transactions of the Royal Society of London A* **266**, 425–533.

Lambert, F. R. and Marshall, A. G. (1991). Keystone characteristics of bird-dispersed *Ficus* in a Malaysian lowland rain forest. *Journal of Ecology* **79**, 793–809.

Lawton, J. H. (1987). Are there assembly rules for successional communities? In Gray, A. J., Crawley, M. J. and Edwards, P. J. (eds.) *Colonization, Succession and Stability*. Oxford, Blackwell Scientific Publications, pp. 225–244.

Lindroth, C. H., Andersson, H., Bödvarsson, H. and Richter, S. H. (1973). Surtsey, Iceland: the development of a new fauna 1963–1970 – terrestrial invertebrates. *Entomologica Scandinavica* **5** (Suppl.), 1–280.

Linsley, E. G. and Gressitt, J. L. (1972). Editorial preface to *Robert Leslie Usinger: Autobiography of an Entomologist*. San Francisco, CA, Pacific Coast Entomological Society.

MacArthur, R. H. (1970). Species packing and competitive equilibrium for many species. *Theoretical Population Biology* **1**, 1–11.

MacArthur, R. H. (1972). *Geographical Ecology*. New York, Harper and Row.

MacArthur, R. H. and Wilson, E. O. (1963). An equilibrium theory of insular zoogeography. *Evolution* **17**, 373–387.

MacArthur, R. H. and Wilson, E. O. (1967). *The Theory of Island Biogeography*. Princeton, NJ, Princeton University Press.

MacMahon, J. A., Parmentier, R. R., Johnson, K. A. and Crisafulli, C. M. (1989). Small mammal recolonization on the Mount St. Helens volcano: 1980–1987. *American Midland Naturalist* **122**, 365–387.

Maeto, K. and Thornton, I. W. B. (1993). A preliminary appraisal of the braconid (Hymenoptera) fauna of the Krakatau Islands (Indonesia) in 1984–1986, with comments on the colonizing abilities of parasitoid modes. *Japanese Journal of Entomology* **61**, 787–801.

Magnússon, B. and Magnússon, S. H. (2000). Vegetation succession on Surtsey, Iceland, during 1990-1998 under the influence of breeding gulls. *Surtsey Research Progress Reports* **12**, 119-120.

Magnússon, B., Magnússon, S. H. and Gudmundsson J. (1996). Gódurframvinder í Surtsey (Vegetation succession on the volcanic island Surtsey). *Búvísindi* **10**, 253-272. (In Icelandic, with English summary)

Maguire, B. Jr (1963). The passive dispersal of small aquatic organisms and their colonization of isolated bodies of water. *Ecological Monographs* **33**, 161-185.

Manning, S. W. (1988). The Bronze Age eruption of Thera: absolute dating, Aegean chronology and Mediterranean cultural interpretation. *Journal of Mediterranean Archaeology* **1**, 17-82.

Manning, S. W. (1989). A new age for Minoan Crete. *New Scientist*, 11 February, 60-63.

Manning, S. W. (1990a). The eruption of Thera: date and implications. In Hardy, D. A. and Renfrew, A. C. (eds.), *Thera and the Aegean World III*, vol. 3, *Chronology*. London, The Thera Foundation, pp. 29-40.

Manning, S. W. (1990b). The Thera eruption: the Third Congress and the problem of the date. *Archaeometry* **32**, 91-100.

Manning, S. W. (1998). Correction. New GISP2 ice-core evidence supports 17th century BC date for the Santorini (Minoan) eruption. *Journal of Archaeological Science* **25**, 1039-1042.

Manning, S. W., Kromer, B., Kuniholm, P. I. and Newton, M. W. (2001). Anatolian tree rings and a new chronology for the East Mediterranean Bronze-Iron Ages. *Science* **294**, 2532-2535.

Manuwal, D. A., Huff, M., Bauer, M., Chappell, C. and Hegstad, K. (1987). Summer birds of the upper subalpine zone of Mount St Helens, Mount Adams and Mount Rainier, Washington. *Northeast Scientist* **61**, 82-92.

Marinatos, S. (1939). The volcanic destruction of Minoan Crete. *Antiquity* **13**, 425-439.

Mazzoleni, S. and Ricciardi, M. (1993). Primary succession on the cone of Vesuvius. In Miles, J. and Walton, D. W. H. (eds.) *Primary Succession on Land*. Oxford, Blackwell Scientific Publications, pp. 101-112.

McKee, C. O., Cooke, R. J. S. and Wallace, D. A. (1976). 1974-75 eruptions of Karkar volcano, Papua New Guinea. In Johnson, R. W. (ed.) *Volcanism in Australasia*. Amsterdam, Elsevier, pp. 173-190.

McKenzie, F., Benton, M. and Hoge, E. J. (1971). *Biological Inventory of the Waters and Keys of Northeast Puerto Rico*, 2nd Report to Division of Natural Resources. San Juan, Commonwealth of Puerto Rico.

McKenzie, N. L., Gunnell, A. C., Yani, M. and Williams, M. R. (1995). Correspondence between flight morphology and foraging ecology in some Palaeotropical bats. *Australian Journal of Zoology* **43**, 241-257.

Mees, G. F. (1986). A list of birds recorded from Bangka Island, Indonesia. *Zoologische Verhandlungen* **232**, 1-176.

Mennis, M. R. (1978). The existence of Yomba Island near Madang: fact or fiction? *Oral History* **6**, 2-81.

Mennis, M. R. (1981). Yomba Island: a real or mythical volcano? *Geological Survey of Papua New Guinea Memoirs* **10**, 115-123.

Menzies, J. L. (1975). *Handbook of Common New Guinea Frogs*. Wau, Papua New Guinea, Wau Ecology Institute.

Michelangeli, F. (2000). Species composition and species-area relationships in vegetation isolates of the Rorairna Tepui. *Journal of Tropical Ecology* **16**, 69-82.

Miles, J. and Walton, D. W. H. (1993). Primary succession revisited. In Miles, J. and Walton, D. W. H. (eds.) *Primary Succession on Land*. Oxford, Blackwell Scientific Publications, pp. 295-302.

Moore, J. G. (1967). Base surges in recent volcanic eruptions. *Bulletin of Volcanology* **30**, 337–363.

Myers, N. (1988). Threatened biotas: 'hot spots' in tropical forests. *Environmentalist* **8**, 187–208.

Myers, N., Mittermeier, R. A., Mittermeier, C. G., da Fonseca, G. A. B. and Kent, J. (2000). Biodiversity hotspots for conservation priorities. *Nature* **403**, 853–858.

Narashimhan, M. J. (1918). Preliminary study of the root nodules of *Casuarina*. *Indian Forester* **44**, 265–268.

New, T. R. and Thornton, I. W. B. (1988). A pre-vegetation population of crickets subsisting on allochthonous aeolian debris on Anak Krakatau. *Philosophical Transactions of the Royal Society of London B* **322**, 481–485.

New, T. R. and Thornton, I. W. B. (1992a). Colonization of the Krakatau Islands by invertebrates. *GeoJournal* **28**, 219–224.

New, T. R. and Thornton, I. W. B. (1992b). The butterflies of Anak Krakatau, Indonesia: faunal development in early succession. *Journal of the Lepidopterists' Society* **46**, 83–96.

New, T. R., Bush, M. B., Thornton, I. W. B. and Sudarman, H. K. (1988). The butterfly fauna of the Krakatau Islands after a century of recolonization. *Philosophical Transactions of the Royal Society of London B* **322**, 445–457.

Newhall, C. G. and Self, S. (1982). The volcanic explosivity index (VEI): an estimate of explosive magnitude for historical volcanism. *Journal of Geophysical Research* **87**, 1231–1238.

O'Brien, T. G. and Kinnaird, M. F. (1996). Changing populations of birds and mammals in North Sulawesi. *Oryx* **30**, 150–156.

Ólafsson, E. (1978). The development of the land-arthropod fauna on Surtsey, Iceland, during 1971–1976, with notes on terrestrial Oligochaeta. *Surtsey Research Progress Reports* **8**, 41–46.

Ólafsson, E. (1982). The status of the land-arthropod fauna on Surtsey, Iceland, in summer 1981. *Surtsey Research Progress Reports* **9**, 68–72.

Osborne, P. I. and Murphy, R. (1989). Botanical colonization of Motmot Island, Lake Wisdom, Madang Province. *Science in New Guinea* **15**, 57–63.

Osman, R. W. (1982). Artificial substrates as ecological islands. In Cairns, J. (ed.) *Artificial Substrates*. Ann Arbor, MI, Ann Arbor Science Publications, pp. 71–114.

Pain, C. F. and Blong, R. J. (1979). The distribution of tephras in the Papua New Guinea highlands. *Search* **10**, 228–230.

Pain, C. F., Blong, R. J. and McKee, C. O. (1981). Pyroclastic deposits and eruptive sequences on Long Island, Papua New Guinea. *Papua New Guinea Geological Survey Memoirs* **10**, 101–107.

Palfreyman, W. D. and Cooke, R. J. S. (1976). Eruptive history of Manam volcano, Papua New Guinea. In Johnson, R. W. (ed.) *Volcanism in Australasia*. Amsterdam, Elsevier, pp. 117–131.

Partomihardjo, T. (1995). Studies on the ecological succession of plants and their associated insects on the Krakatau Islands, Indonesia. D.Phil. thesis, University of Kagoshima, Japan.

Partomihardjo, T. (1997). Flora of Sebesi Island: its role and potential in the recolonization process of the Krakatau Islands. *Seminar Nasional Konservasi Flora Nusantara*, **1997**, 15–19.

Partomihardjo, T., Mirmanto, E. and Whittaker, R. J. (1992). Anak Krakatau's vegetation and flora circa 1991, with observations on a decade of development and change. *GeoJournal* **28**, 233–248.

Partomihardjo, T., Mirmanto, E., Riswan, S. and Suzuki, E. (1993). Drift fruit and seeds on Anak Krakatau beaches, Indonesia. *Tropics* **2**, 143–156.

Patrick, R. (1967). The effect of invasion rate, species pool, and size of area on the structure of the diatom community. *Proceedings of the National Academy of Sciences of the USA* **58**, 1335–1342.

Patrick, R. (1968). The structure of diatom communities in similar ecological conditions. *American Naturalist* **102**, 173–183.

Peck, S. (1996). Origin and development of an insect fauna on a remote tropical archipelago: the Galapagos Islands, Ecuador. In Keast, A. and Miller, S. E. (eds.) *The Origin and Evolution of Pacific Island Biotas, New Guinea to Eastern Polynesia: Patterns and Processes.* Amsterdam, SPB Academic Publishers, pp. 91–121.

Pendick, D. (1996). Return to Mount St. Helens. *Earth* **4**, 24–33.

Polach, H. A. (1981). Pyroclastic deposits and eruptive sequences of Long Island. II. Radiocarbon dating of Long Island and Tibito tephras. *Geological Survey of Papua New Guinea Memoirs* **10**, 108–113.

Poonamperuma, C., Young, R. S. and Caren, L. D. (1967). Some chemical and microbiological studies of Surtsey. *Surtsey Research Progress Reports* **3**, 70–80.

Post, W. M. and Pimm, S. L. (1983). Community assembly and food web stability. *Mathematical Biosciences* **64**, 169–192.

Pratt, W. E. (1911). The eruption of Taal Volcano, 30 January 1911. *Philippine Journal of Science, Series A* **6**, 63–83.

Preston, F. W. (1962). The canonical distribution of commonness and rarity. *Ecology* **43**, 185–215, 410–432.

Pyke, D. A. (1984). Initial effect of volcanic ash from Mount St Helens on *Peromyscus maniculatus* and *Microtus montanus*. *Journal of Mammalogy* **65**, 678–680.

Pyle, D. M. (1990a). New estimates for the volume of the Minoan eruption. In Hardy, D. A., Keller, J., Galanopoulos, V. P., Flemming, N. C. and Druitt, T. H. (eds.) *Thera and the Aegean World III*, vol. 2, *Earth Sciences.* London, The Thera Foundation, pp. 113–121.

Pyle, D. M. (1990b). The application of tree-ring and ice-core studies to the dating of the Minoan eruption. In Hardy, D. A. and Renfrew, A. C. (eds.) *Thera and the Aegean World III*, vol. 3, *Chronology.* London, The Thera Foundation, pp. 167–173.

Pyle, R. M. (1984). The impact of recent vulcanism on Lepidoptera. In Vane-Wright, R. I. and Ackery, P. R. (eds.) *The Biology of Butterflies.* London, Academic Press, pp. 323–326.

Raab, T. K., Lipson, D. A. and Monson, R. K. (1999). Soil amino acid utilization among species of the Cyperaceae: plant and soil processes. *Ecology* **80**, 2408–2419.

Rackham, O. (1978). The flora and vegetation of Thera and Crete before and after the great eruption. In Doumas, C. (ed.) *Thera and the Aegean World*, vol. 1. London, The Thera Foundation, pp. 755–764.

Rackham, O. (1990). Observations on the historical ecology of Santorini. In Hardy, D. A., Keller, J., Galanopoulos, V. P., Flemming, N. C. and Druitt, T. H. (eds.) *Thera and the Aegean World III*, vol. 2 *Earth Sciences.* London, The Thera Foundation, pp. 384–391.

Rampino, M. R. and Self, S. (1982). Historic eruptions of Tambora (1815), Krakatau (1883) and Agung (1963), their stratospheric aerosols and climatic impact. *Quaternary Research* **18**, 127–143.

Rampino, M. R. and Self, S. (1984). Sulphur-rich volcanic eruptions and stratospheric aerosols. *Nature* **310**, 677–679.

Raus, T. (1986). Floren- und Vegetationsdynamik auf der Vulcaninsel Nea Kameni (Santorin-Archipel, Kykladen, Griechenland). *Abhandlung der*

Landesmuseum für Naturkunde Münster Westfalen **48**, 373–394.

Raus, T. (1988). Vascular plant colonization and vegetation development on sea-born volcanic islands in the Aegean (Greece). *Vegetatio* **77**, 139–147.

Raus, T. (1991). Die Flora (Farne und Blutenpflanzen) des Santorin-Archipels. In Schmalfuss, H. (ed.) *Santorin: Leben auf Schutt und Asche*. Weikersheim, Verlag J. Margaraf, pp. 109–124.

Rawlinson, P. A., Widjoya, A. H. T., Hutchinson, M. N. and Brown, G. W. (1990). The terrestrial vertebrate fauna of the Krakatau Islands, Sunda Strait, 1883–1986. *Philosophical Transactions of the Royal Society of London B* **328**, 3–28.

Rawlinson, P. A., Zann, R. A., van Balen, S. and Thornton, I. W. B. (1992). Colonization of the Krakatau Islands by vertebrates. *GeoJournal* **28**, 225–231.

Recher, H. F. and Serventy, D. L. (1991). Long-term changes in the relative abundance of birds in Kings Park, Perth, Western Australia. *Conservation Biology* **5**, 90–120.

Rechinger, K. (1910). Botanische und zoologische Ergebnisse einer Forschungsreise nach den Sampa-inseln, dem Neuguinea-archipel und den Salomoninseln. III. Siphonogamen der Samoa-inseln. *Denkschriften der Akademie der Wissenschaften Wien* **85**, 202–388.

Renfrew, A. C. (1990a.) Introductory remarks. In Hardy, D. A. and Renfrew, A. C. (eds.) *Thera and the Aegean World III*, vol. 3, *Chronology*. London, The Thera Foundation, pp. 11–13.

Renfrew, A. C. (1990b.) Summary of the progress in chronology. In Hardy, D. A. and Renfrew, A. C. (eds.) *Thera and the Aegean World III*, vol. 3, *Chronology*. London, The Thera Foundation, pp. 342.

Renfrew, C. (1996). Kings, tree rings and the Old World. *Nature* **381**, 733–734.

Reynolds, M. A. and Best, J. G. (1976). Survey of the 1953–57 eruption of Tuluman Volcano, Papua New Guinea. In Johnson, R. W. (ed.) *Volcanism in Australasia*. Amsterdam, Elsevier, pp. 287–296.

Richards, P. W. (1952). *Tropical Rainforest: An Ecological Study*. Cambridge, Cambridge University Press.

Robinson, J. V. and Dickerson, J. E. Jr (1987). Does invasion sequence affect community structure? *Ecology* **68**, 587–589.

Robinson, J. V. and Edgemon, M. A. (1988). An experimental evaluation of the effect of invasion history on community structure. *Ecology* **69**, 1410–1417.

Rosenzweig, M. L. (1995). *Species Diversity in Space and Time*. Cambridge, Cambridge University Press.

Roughgarden, J. (1989). The structure and assembly of communities. In Roughgarden, J., May, R. M. and Levin, S. A. (eds.) *Perspectives in Ecological Theory*. Princeton, NJ, Princeton University Press, pp. 203–226.

Runciman, D., Cook, S., Riley, J. Wardell, J. and Thornton, I. W. B. (1998). The avifauna of Sebesi, a possible stepping stone to the Krakatau Islands. *Tropical Biodiversity* **5**, 1–9.

Sands, W. N. (1912). An account of the return of vegetation and the revival of agriculture in the area devastated by the Soufriere of St Vincent in 1902/3. *West Indian Bulletin* **12**, 22–23.

Schedvin, N., Cook, S. and Thornton, I. W. B. (1995). The diversity of bats on the Krakatau Islands, Indonesia. *Biodiversity Letters* **2**, 87–92.

Schipper, C., Shanahan, M., Cook, S. and Thornton, I. W. B. (2001). Colonization of an island volcano, Long Island, Papua New Guinea, and of an emergent island, Motmot, in its caldera lake. III.

Colonization by birds. *Journal of Biogeography* **28**, 1339–1352.

Schmalfuss, H. and Schawaller, W. (1984). Die Fauna der Ägäis-Insel Santorin. V. Arachnida und Crustacea. *Stuttgarter Beiträge Naturkunde, Serie A, Biologie* **371**, 1–16.

Schmalfuss, H., Steidel, C. and Schlegel, M. (1981). Die Fauna der Ägäis-Insel Santorin. I. *Stuttgarter Beiträge Naturkunde, Serie A, Biologie* **347**, 1–14.

Schmidt, E. R., Thornton, I. W. B. and Hancock, D. (1994). Tropical fruitflies (Diptera: Tephritidae) of the Krakatau Archipelago in 1990 and comments on faunistic changes since 1982. *Ecological Research* **9**, 1–8.

Schmitt, S. F. and Partomihardjo, T. (1997). Disturbance and its significance for forest succession and diversification on the Krakatau Islands, Indonesia. In Dransfield, J., Coode, M. J. E. and Simpson, D. A. (eds.) *Plant Diversity in Malesia III*. Kew, Royal Botanic Gardens, pp. 247–263.

Schoener, A. (1988). Experimental island biogeography. In Myers, A. A. and Giller, P. S. (eds.) *Analytical Biogeography: An Integrated Approach to the Study of Animal and Plant Distributions*. London, Chapman and Hall, pp. 483–512.

Schwabe, G. H. (1971). Die Ökogenese im terrestrichen Bereich postvulcanische Substrate, Schematische Obersicht bischeriger Befunde auf Surtsey, Iceland. *Petermanns Geographische Mitteilungen* **4**, 168–173.

Sear, C. B., Kelley, P. M., Jones, P. D. and Goodess, C. M. (1987). Global surface temperature responses to major volcanic eruptions. *Nature* **330**, 365–367.

Selenka, E. and Selenka, K. (1905). *Sonnige Welten*, 2nd edn. Wiesbaden, C. W. Kneidels Verlag.

Shanahan, M. and Compton, S. G. (2001). Vertical stratification of figs and fig eaters in a Bornean lowland rainforest: how is the canopy different? *Plant Ecology* **153**, 121–132.

Shanahan, M., Harrison, R. D., Yamuna, R. and Thornton, I. W. B. (2001). Colonization of an island volcano, Long Island, Papua New Guinea, and of an emergent island, Motmot, in its caldera lake. V. Colonization by figs (*Ficus* species), their dispersers and pollinators. *Journal of Biogeography* **28**, 1365–1377.

Shilton, L. A., Altringham, J. D., Compton, S. G. and Whittaker, R. J. (1999). Old World fruit bats can be long-distance seed dispersers through extended retention of viable seed in their gut. *Proceedings of the Royal Society of London B* **266**, 219–223.

Shiro, T. (1991). Species turnover and diversity during early stages of vegetation recovery on the volcano Usu, northern Japan. *Journal of Vegetation Science* **2**, 301–306.

Shiro, T. and del Moral, R. (1993). Species attributes in early primary succession on volcanoes. *Journal of Vegetation Science* **6**, 517–522.

Shmida, A. and Ellner, S. (1984). Coexistence of plant species with similar niches. *Vegetatio* **58**, 29–55.

Sigurdardóttir, H. (2000). Status of collembolans (Collembola) on Surtsey, Iceland in 1995 and first encounter of earthworms (Lumbricidae) in 1993. *Surtsey Research* **11**, 51–55.

Silvester, W. B. (1977). Dinitrogen fixation by plant associations excluding legumes. In Hardy, R. W. F. and Gibson, A. H. (eds.) *A Treatise on Dihydrogen Fixation*. New York, John Wiley, pp. 141–190.

Simberloff, D. (1976). Species turnover and equilibrium island biogeography. *Ecology* **57**, 629–648.

Simberloff, D. and Wilson, E. O. (1969). Experimental zoogeography of islands: the colonization of empty islands. *Ecology* **50**, 278–296.

Simberloff, D. and Wilson, E. O. (1970). Experimental zoogeography of islands: a two-year record of colonization. *Ecology* **51**, 934–937.

Sipman, H. J. M. and Raus, T. (1995). Lichen observations from Santorini (Greece). *Bibliotheca Lichenologica* **57**, 409–428.

Smith, B. J. and Djajasasmita, M. (1988). The land molluscs of the Krakatau Islands, Indonesia. *Philosophical Transactions of the Royal Society of London B* **323**, 379–400.

Smith, J. D. and Hood, C. S. (1981). Preliminary notes on bats from the Bismarck Archipelago (Mammalia: Chiroptera). *Science in New Guinea* **8**, 81–121.

Sohlenius, B. (1974). Nematodes from Surtsey. II. *Surtsey Research Progress Reports* **7**, 35.

Sparks, R. S. J. and Wilson, C. J. N. (1990). The Minoan deposits: a review of their characteristics and interpretation. In Hardy, D. A., Doumas, C. G., Sakellarakis, J. A. and Warren, P. M. (eds.) *Thera and the Aegean World III*, vol. 1, *Archaeology*. London, The Thera Foundation, pp. 89–98.

Specht, J., Ball, E. E, Blong, R. J. *et al.* (1982) Long Island, Papua New Guinea: introduction. *Records of the Australian Museum* **34**, 407–417.

Sprent, J. I. (1993). The role of nitrogen fixation in primary succession on land. In Miles, J. and Walton, D. W. H. (eds.) *Primary Succession on Land*. Oxford, Blackwell Scientific Publications, pp. 209–220.

Sugg, P. M. (1986). Arthropod populations at Mount St. Helens: survival and revival. In Keller, S. A. C. (ed.) *Mount St. Helens: Five Years Later*. Spokane, WA, Eastern Washington University Press, pp. 325–328.

Sugg, P. M. and Edwards, J. S. (1998). Pioneer Aeolian community development on pyroclastic flows after the eruption of Mount St Helens. *Arctic and Alpine Research* **30**, 400–407.

Sugg, P. M., Greve, L. and Edwards, J. S. (1994). Neuropteroidea from Mount St. Helens and Mount Rainier: dispersal and immigration in volcanic landscapes. *Pan-Pacific Entomologist* **70**, 212–221.

Sullivan, D. G. (1988). The discovery of Santorini Minoan tephra in western Turkey. *Nature* **333**, 552–554.

Surtsey (2006). *Bibliography of Scientific Research*. Available online at www.surtsey.is/

Swanson, F. J. (1987). Ecological effects of the eruption of Mount St. Helens: an overview. In Bilderback, D. E. (ed.) *Mount St. Helens 1980: Botanical Consequences of the Explosive Eruption*. Berkeley, CA, University of California Press, pp. 1–2.

Tagawa, H. (1992). Primary succession and the effect of first arrivals on the subsequent development of forest types. *GeoJournal* **28**, 175–183.

Tagawa, H. (2005). *The Krakataus: Changes in a Century since Catastrophic Eruption in 1883*. Kagoshima, University of Kagoshima.

Tagawa, H., Suzuki, E., Partomihardjo, T. and Suriadarma, A. (1985). Vegetation and succession on the Krakatau Islands, Indonesia. *Vegetatio* **60**, 131–145.

Talling, J. F. (1951). The element of chance in pond populations. *The Naturalist*, October–December, 157–170.

Taylor, G. A. M. (1953). Seismic and tilt phenomena preceding a Pelean type eruption from a basaltic volcano. *Bulletin of Volcanology* **26**, 5–11.

Thorarinsson, S. (1971). Damage caused by the tephra fall in some big Icelandic eruptions and its relation to the thickness of the tephra layers. *Proceedings* of the International Scientific Congress on the Volcano of Thera, Athens, pp. 213–236.

Thorarinsson, S. (1978). Some comments on the Minoan eruption of Santorini. In Thera and the Aegean World I, Papers presented at the 2nd International Scientific Congress, Antorini, Greece, August 1978. London, Thera and the Aegean World, pp. 263–276.

Thornton, I. (1971). Darwin's Islands: A Natural History of the Galapagos. New York, Doubleday.

Thornton, I. W. B. (1991). Krakatau: studies on the origin and development of a fauna. In Dudley, E. C. (ed.) The Unity of Evolutionary Biology. Portland, OR, Dioscorides Press, pp. 396–408.

Thornton, I. W. B. (ed.) (1992a). Krakatau: a century of change. GeoJournal 28, 81–302.

Thornton, I. W. B. (1992b). K. W. Dammerman: fore-runner of island equilibrium theory? Global Ecology and Biogeography Letters 2, 145–148.

Thornton, I. W. B. (1994) Figs, frugivores and falcons: an aspect of the assembly of mixed tropical forest on the emergent volcanic island Anak Krakatau. South Australian Geographical Journal 93, 3–21.

Thornton, I. (1996a). Krakatau: The Destruction and Reassembly of an Island Ecosystem. Cambridge, MA, Harvard University Press.

Thornton, I. W. B. (1996b). The origins and development of island biotas as illustrated by Krakatau. In Keast, A. and Miller, S. G. (eds.) The Origin and Evolution of Pacific Island Biotas, New Guinea to Eastern Polynesia: Patterns and Processes. Amsterdam, SPB Academic Publishing, pp. 67–90.

Thornton, I. W. B. (2001). Colonization of an island volcano, Long Island, Papua New Guinea, and of an emergent island, Motmot, in its caldera lake. I. General

introduction. Journal of Biogeography 28, 1299–1310.

Thornton, I. W. B. and New, T. R. (1988a). Freshwater communities on the Krakatau Islands. Philosophical Transactions of the Royal Society of London B 322, 487–492.

Thornton, I. W. B. and New, T. R. (1988b). Krakatau invertebrates: the 1980s fauna in the context of a century of colonization. Philosophical Transactions of the Royal Society of London B 322, 493–522.

Thornton, I. W. B. and Walsh, D. (1992). Photographic evidence of rate of development of plant cover on the emergent island Anak Krakatau from 1971 to 1991 and implications for the effect of volcanism. GeoJournal 28, 249–259.

Thornton, I. W. B., New, T. R., McLaren, D. A., Sudarman, H. K. and Vaughan, P. J. (1988). Air-borne arthropod fall-out on Anak Krakatau and a possible pre-vegetation pioneer community. Philosophical Transactions of the Royal Society of London B 322, 471–479.

Thornton, I. W. B., Zann, R. A., Rawlinson, P. A. et al. (1989). Colonization of the Krakatau Islands by vertebrates: equilibrium, succession, and possible delayed extinction. Proceedings of the National Academy of Sciences of the USA 85, 515–518.

Thornton, I. W. B., Zann, R. A. and Stephenson, D. G. (1990a). Colonization of the Krakatau Islands by land birds, and the approach to an equilibrium number of species. Philosophical Transactions of the Royal Society of London B 328, 55–93.

Thornton, I. W. B., New, T. R., Zann, R. A. and Rawlinson, P. A. (1990b). Colonization of the Krakatau Islands by animals: a perspective from the 1980s. Philosophical Transactions of the Royal Society of London B 328, 132–165.

Thornton, I. W. B., Ward, S. A., Zann, R. A. and New, T. R. (1992). Anak Krakatau: a colonization model within a colonization model? *GeoJournal* **28**, 271–286.

Thornton, I. W. B., Zann, R. A. and van Balen, S. (1993). Colonization of Rakata (Krakatau Is.) by non-migrant land birds from 1883 to 1992 and implications for the value of island equilibrium theory. *Journal of Biogeography* **20**, 441–452.

Thornton, I. W. B., Compton, S. G. and Wilson, C. N. (1996). The role of animals in the colonization of the Krakatau Islands by fig trees (*Ficus* species). *Journal of Biogeography* **23**, 577–592.

Thornton, I. W. B., Mawdsley, N. A. and Partomihardjo, T. (2000). Persistence of biota on Anak Krakatau after a three-year period of volcanic activity. *Tropical Biodiversity* **7**, 25–43.

Thornton, I. W. B., Cook, S., Edwards, J. S. *et al.* (2001). Colonization of an island volcano, Long Island, Papua New Guinea, and of an emergent island, Motmot, in its caldera lake. VII. Overview and discussion. *Journal of Biogeography* **28**, 1389–1408.

Thornton, I. W. B., Runciman, D., Cook, S. *et al.* (2002). How important were stepping stones in the colonization of Krakatau? *Biological Journal of the Linnean Society* **77**, 275–317.

Tidemann, C. R., Kitchener, D. J., Zann, R. A. and Thornton, I. W. B. (1990). Recolonization of the Krakatau Islands and adjacent areas of West Java, Indonesia, by bats (Chiroptera) 1883–1886. *Philosophical Transactions of the Royal Society of London B* **328**, 123–130.

Toft, C. A. and Schoener, T. W. (1983). Abundance and diversity of orb spiders on 106 Bahamian islands: biogeography at an intermediate trophic level. *Oikos* **41**, 411–426.

Torrey, J. G. (1983). *Casuarina*: actinorhizal dinitrogen-fixing tree of the tropics. In Midgley, S. J., Turnbull, J. W. and Johnston, R. D. (eds.) *Casuarina Ecology Management and Utilization*. Canberra, CSIRO, pp. 193–204.

Toxopeus, L. J. (1950). Over de pioneer-fauna van Anak Krakatau, met enige beschouwingen over het onstaat van de Krakatau-fauna. *Chronica Naturae* **106**, 27–34.

Trantalidou, C. (1990). Animals and human diet in the prehistoric Aegean. In Hardy, D. A., Keller, J., Galanopoulos, V. P., Flemming, N. C. and Druitt, T. H. (eds.) *Thera and the Aegean World III*, vol. 2, *Earth Sciences*. London, The Thera Foundation, pp. 392–405.

Treub, M. (1888). Notice sur la nouvelle flore de Krakatau. *Annales du Jardin Botanique de Buitenzorg* **7**, 213–223.

Turner, B. (1992). The colonization of Anak Krakatau: interactions between wild sugar cane, *Saccharum spontaneum*, and the ant lion, *Myrmeleon frontalis*. *Journal of Tropical Ecology* **8**, 435–449.

Turner, B. D. (1997). Patterns of change in arthropod biodiversity living on *Casuarina equisetifolia*, an early successional tree species on the island of Anak Krakatau, Indonesia. *Tropical Biodiversity* **4**, 241–257.

Underwood, A. J., Denley, E. J. and Moran, M. J. (1983). Experimental analysis of the structure and dynamics of mid-shore intertidal communities in New South Wales. *Oecologia* **56**, 202–219.

van Bemmelen, R. W. (1971). Four volcanic outbursts that influenced human history: Toba, Sunda, Merapi, and Thera. *Proceedings of the International Scientific Conference on the Volcano of Thera*, Athens, pp. 5–50.

van Borsum Waalkes, J. (1954). The Krakatau Islands after the eruption of October 1952. *Penggemar Alam* **34**, 97–104.

van Borssum Waalkes, J. (1960). Botanical observations on the Krakatau Islands in 1951 and 1952. *Annales Bogoriensis* **4**, 5–63.

van Tol, J. (1990). Zoological expeditions to the Krakataus 1984 and 1985. Odonata. *Tijdschrifte vor Entomologie* **133**, 273–279.

Vaupel, F. (1910). Die Vegetation der Samoa-Inseln. *Botanisches Jahrbuch* **44**, 47–58.

Verbeek, R. D. M. (1884). The Krakatoa eruption. *Nature* **30**: 10–15.

Verbeek, R. D. M. (1885). *Krakatau*. Batavia, Landsdrukkerij.

Waide, R. B. (1991). Summary of the responses of animal populations to hurricanes in the Caribbean. *Biotropica* **23**, 508–512.

Waldron, H. H. (1967). *Debris Flow and Erosion Control Problems Caused by the Ash Eruptions of Irazu Volcano, Costa Rica*, US Geological Survey Bulletin no. 1241-1. Washington, DC, Government Printing Office.

Ward, S. A. and Thornton, I. W. B. (1998). Guest editorial: equilibrium theory and alternative stable equilibria. *Journal of Biogeography* **25**, 615–622.

Ward, S. A. and Thornton, I. W. B. (2000). Chance and determinism in the development of isolated communities. *Global Ecology and Biogeography Letters* **9**, 7–18.

Warren, P. (1984). Absolute dating of the Bronze Age eruption of Thera (Santorini). *Nature* **305**, 492–493.

Waters, A. C. and Fisher, R. V. (1971). Base surges and their deposits: Capelhinos and Taal volcanoes. *Journal of Geophysical Research* **76**, 5595–5614.

Watkins, N. D., Sparks, R. S. J., Sigurdsson, H. *et al.* (1978). Volume and extent of the Minoan tephra from Santorini Volcano: new evidence from deep-sea sediment cores. *Nature* **271**, 122–126.

Whittaker, R. J. (1998). *Island Biogeography: Ecology, Evolution and Conservation*. Oxford, Oxford University Press.

Whittaker, R. J. and Jones, S. H. (1994). The role of frugivorous bats and birds in the rebuilding of a tropical forest ecosystem, Krakatoa, Indonesia. *Journal of Biogeography* **21**, 246–258.

Whittaker, R. J., Bush, M. B. and Richards, K. (1989). Plant recolonization and vegetation succession on the Krakatau Islands, Indonesia. *Ecological Monographs* **59**, 59–123.

Whittaker, R. J., Bush, M. B., Partomihardjo, T., Asquith, N. M. and Richards, K. (1992a). Ecological aspects of plant colonization of the Krakatau Islands. *GeoJournal* **28**, 201–211.

Whittaker, R. J., Walden, J. and Hill, J. (1992b). Post-1883 ash fall on Panjang and Sertung and its ecological impact. *GeoJournal* **28**, 153–171.

Whittaker, R. J., Partomihardjo, T. and Riswan, S. (1995). Surface and buried seed banks from Krakatau, Indonesia: implications for the sterilization hypothesis. *Biotropica* **27**, 345–354.

Whittaker, R. J., Jones, S. H. and Partomaihardjo, T. (1997). The re-building of an isolated rain forest assembly; how disharmonious is the flora of Krakatau? *Biodiversity and Conservation* **6**, 1671–1696.

Whittaker, R. J., Field, R. and Partomihardjo, T. (2000). How to go extinct: lessons from the lost plants of Krakatau. *Journal of Biogeography* **27**, 1049–1064.

Williamson, M. (1982). *Island Populations*. Oxford, Oxford University Press.

Wilson, E. O. (1992). *The Diversity of Life*. Cambridge, MA, Harvard University Press.

Winchester, S. (2003). *Krakatoa: The Day the World Exploded – 27th August 1883*. London, Viking.

Winoto Suatmadji, R., Coomans, A., Rashid, F., Gevaert, E. and McLaren, D. A. (1988). Nematodes of the Krakatau archipelago, Indonesia: a preliminary overview.

Philosophical Transactions of the Royal Society of London B **322**, 369–378.

Wissel, C. and Maier, B. (1992). A stochastic model for the species-area relationship. *Journal of Biogeography* **19**, 355–362.

Wurmli, M. (1974). Biocenoses and their successions on the lava and ash of Mt. Etna, Part 1. *Image Roche* **59**, 32–40.

Yamane, S. (1988). The aculeate fauna of the Krakatau Islands (Insecta, Hymenoptera). *Reports of the Faculty of Science, Kagoshima University (Earth Sciences and Biology)* **16**, 75–107.

Yamane, S. and Tomiyama, K. (1986). A small collection of land snails from the Krakatoa Islands, Indonesia. *Venus*, **45**, 61–64.

Yamane, S., Abe, T. and Yukawa, J. (1992). Recolonization of the Krakataus by Hymenoptera and Isoptera (Insecta). *GeoJournal* **28**, 213–218.

Yih, K., Boucher, D. H., Vandermeer, J. H. and Zamora, N. (1991). Recovery of the rain forest of southeastern Nicaragua after destruction by Hurricane Joan. *Biotropica* **23**, 106–113.

Yokoyama, I. (1978). The tsunami generated by the prehistoric eruption of Thera. In *Thera and the Aegean World I, Papers presented at the 2nd International Scientific Congress, Antorini, Greece, August 1978.* London, Thera and the Aegean World, pp. 277–286.

Yukawa, J. (1984). Geographical ecology of the butterfly fauna of the Krakatau Islands, Indonesia. *Tyô to Ga* **36**, 181–184.

Yukawa, J. and Yamane, S. (1985). Odonata and Hemiptera collected from the Krakataus and the surrounding islands, Indonesia. *Kontyû* **53**, 690–698.

Yukawa, J., Abe, T., Iwamoto, T. and Yamane, S. (1984). The fauna of Krakatau, Peucang and Panaitan islands. In Tagawa, H. (ed.) *Researches on the Ecological Succession and the Formation Process of Volcanic Ash Soils on the Krakatau Islands*. Kagoshima, Kagoshima University, pp. 91–114.

Yukawa, J., Partomihardjo, T., Yata, O. and Hirowatari, T. (2000). An assessment of the role of Sebesi Island as a stepping stone for the colonization of the Krakatau islands by butterflies. *Esakia* **40**, 1–10.

Zann, R. A. and Darjono. (1992). The birds of Anak Krakatau: the assembly of an avian community. *GeoJournal* **28**, 261–270.

Zann, R. A., Male, E. B. and Darjono. (1990). Bird colonization of Anak Krakatau, an emergent volcanic island. *Philosophical Transactions of the Royal Society of London B* **328**, 95–121.

Z'graggen, J. A. (1975). *The Languages of the Madang District, Papua New Guinea*, Pacific Linguistics Series B, No. 41. Canberra, Department of Linguistics, Research School of Pacific Studies, Australian National University.

Zielinski, G. A. and Germani, M. S. (1998a). New ice-core evidence challenges the 1620s BC age for the Santorini (Minoan) eruption. *Journal of Archaeological Science* **25**, 279–289.

Zielinski, G. A. and Germani, M. S. (1998b). Reply to: Correction. New GISP2 Ice-Core evidence supports 17th century BC date for the Santorini (Minoan) eruption. *Journal of Archaeological Science* **25**, 1043–1045.

Zielinski, G. A., Mayewski, P. A., Meeker, L. D. *et al.* (1994). Record of volcanism since 7000 BC from the GISP2 Greenland Ice Core and implications for the volcano-climate system. *Science* **264**, 948–951.

Zielinski, G. A., Mayewski, P. A., Meeker, L. D., Whitlow, S. and Twickler, M. (1996). A 110 000-year record of explosive volcanism from the GISP2 (Greenland) ice core. *Quaternary Research* **45**, 109–118.

Zimmerman, J. K., Willig, M. R., Walker, L. R. and Silver, W. L. (1996). Introduction, disturbance and Caribbean ecosystems. *Biotropica* **28**, 414–423.

Index